The Organization of Global Negotiations: Constructing the Climate Change Regime

The Organization of Global Negotiations: Constructing the Climate Change Regime

Joanna Depledge

Routledge
Taylor & Francis Group

LONDON AND NEW YORK

earthscan
from Routledge

First published 2005 by Earthscan

Published 2022 by Routledge
2 Park Square, Milton Park, Abingdon, Oxon OX14 4RN
605 Third Avenue, New York, NY 10017

Routledge is an imprint of the Taylor & Francis Group, an informa business

Publisher's Note

The publisher has gone to great lengths to ensure the quality of this
reprint but points out that some imperfections in the original copies
may be apparent.

Typesetting by FiSH Books

Cover design by Danny Gillespie

A catalogue record for this book is available from the British Library

Library of Congress Cataloging-in-Publication Data

Depledge, Joanna.
The organization of international negotiations: constructing the climate
change regime/Joanna Depledge.
   p. cm.

   Includes bibliographical references and index.
   ISBN 1-84407-046-8 (hardback)
   1. Climatic changes – International cooperation. 2. Environmental policy –
International cooperation. I. Title
   QC981.8.C5D43 2005
   363.738'75526–dc22

   2004024.817

   ISBN13: 978-1-84407-046-6 (hbk)
   ISBN13: 978-1-138-97769-3 (pbk)

Welcome to the Maldives Climate Conference

Source: © Lawrence Moore

Contents

List of Figures, Tables and Boxes

Figures

Tables

Boxes

Acknowledgements

This book began life as a PhD thesis. First and foremost, therefore, thanks are due to my supervisor, Professor Jacquie Burgess, at University College London. I am sincerely grateful for her unfailing support throughout my PhD process, which was a somewhat unusual one. Without Jacquie's encouragement, my thesis would never have metamorphosed into this book. Thanks are also due to the UK Economic and Social Research Council (ESRC) for their invaluable financial support (postgraduate training award R00429634040), and for having demonstrated the administrative flexibility necessary to accommodate my work at the UNFCCC secretariat in Bonn.

I am deeply indebted to all the interviewees who took time out from the climate change negotiations to share their experiences with me and who, in doing so, have greatly enriched this book. It would exceed the page limit of this book to express the appreciation deserved by all my climate change friends and colleagues around the world who have inspired, encouraged, informed and enlightened me over the past years. I must, however, single out five individuals, who laid the foundations for this book: Richard Kinley, Michael Zammit Cutajar and Greg Terrill, who understand the importance of effective organization; Ambassador Estrada, without whom the story of this book would have been very different; and Dr Ian Rowlands, formerly at the London School of Economics and Political Science, now at the University of Waterloo in Canada, who instilled in me a fascination with climate change that has only grown over the years.

And finally, a big thank you to everyone at Lucy Cavendish College and the Sutasoma Trust for having faith in my research, and to Jonathan Sinclair Wilson, Ruth Mayo, Camille Adamson and their colleagues at Earthscan for supporting me throughout this project, and for greeting my countless requests for more time with good cheer and understanding.

Acronyms and Abbreviations

AIJ	activities implemented jointly
AG13	Ad Hoc Group on Article 13
AGBM	Ad Hoc Group on the Berlin Mandate
AOSIS	Alliance of Small Island States
BAPA	Buenos Aires Plan of Action (agreed at COP 4 in 1998)
BINGO	business and industry non-governmental organization
CBD	Convention on Biological Diversity (1992)
CDM	clean development mechanism
COP/MOP	Conference of the Parties serving as the meeting of the Parties to the Kyoto Protocol
COP	Conference of the Parties
CoW	Committee of the Whole (at COP 3)
CRP	conference room paper
EIT	economy in transition (former Soviet Union and Eastern Europe)
ENB	Earth Negotiations Bulletin
ENGO	environmental non-governmental organization
G-77	Group of 77
GEF	Global Environment Facility
GHG	greenhouse gas
IGO	intergovernmental organization
IHLP	informal high-level plenary (at COP 6)
INC	Intergovernmental Negotiating Committee for the UNFCCC (1990–1995)
INF	Information document
IPCC	Intergovernmental Panel on Climate Change
IPOs	Indigenous People's Organizations
JI	joint implementation
JWG	Joint Working Group on Compliance
LDC	least developed country
LGMAs	Local Government and Municipal Authorities
LULUCF	land use, land-use change and forestry
MEA	multilateral environmental agreement
MISC	miscellaneous document
NGO	non-governmental organization
OECD	Organisation for Economic Co-operation and Development
OPEC	Organization of Petroleum Exporting Countries
QELROs	quantified emission limitation and reduction objectives
RINGO	research-oriented and independent non-governmental organizations

SBI	Subsidiary Body for Implementation
SBSTA	Subsidiary Body for Scientific and Technological Advice
SIDS	Small Island Developing States
TAR	Third Assessment Report of the IPCC
UNCTAD	United Nations Conference on Trade and Development
UNEP	United Nations Environment Programme
UNFCCC	United Nations Framework Convention on Climate Change
UNGA	United Nations General Assembly
UNOG	United Nations Office at Geneva
WMO	World Meteorological Organization
WTO	World Trade Organization

1

Introduction

The balance between war and peace may be a matter not of the
nature of the differences that divide us, but of the *process we use
to resolve those differences* (Raiffa, 1991, p9).

Global negotiations – negotiations that are open to all of the world's nation
states – have become an increasingly popular means of tackling pressing
problems that cut across international boundaries. Environmental issues have
been at the forefront of this trend, with global negotiations at the close of the
last millennium agreeing treaties on, for example, climate change, biodiversity
loss, desertification, persistent organic pollutants, prior informed consent and
stratospheric ozone depletion. Other issues have also been addressed through
global negotiations, including anti-personnel landmines through the 1997
Ottawa Convention, international trade under the 1994 World Trade
Organization (WTO) and even smoking under the 2003 World Health
Organization Framework Convention on Tobacco Control, the first public
health global treaty. The focus of this book is on the global negotiations
attempting to resolve one of the more complex and difficult issues facing the
international community – global climate change.

Organizing global negotiations – among more than 180 heterogeneous
states on often highly contentious issues to forge a mutually acceptable outcome
– is a difficult and intricate task. Organizational factors, however, such as the
role of the Chair, the choice of negotiating arenas, the rules for the conduct of
business and the approach of negotiating texts, are usually taken for granted and
rarely attract attention until something goes wrong. A series of high profile
collapses in negotiations – the first round of negotiations on the Cartagena
Biosafety Protocol (February, 1999) the Sixth Conference of the Parties (COP
6) to the United Nations Framework Convention on Climate Change
(UNFCCC) (November, 2000) and the Seattle and Cancún Ministerial
Conferences of the WTO (December, 1999 and December, 2003) – have
gradually begun to draw attention to the dangers of ineffective organization, and
how this can contribute to unnecessary negotiating failure.

The basic assumption of this book is that *the organization of a negotiation
process matters*. The way in which a negotiation is organized can impact on the
negotiation process in a positive or negative way to enhance or reduce its
effectiveness and the likelihood of reaching agreement. This basic assumption is

supported by both participants in international negotiations and the academic literature. Lang, for example, a veteran negotiator in global environmental regimes, states that 'organizational features of a specific negotiation constitute ... important background conditions which are able to facilitate or delay progress' (Lang, 1994, p206). Boyer, in turn, introducing a volume of the leading academic journal *International Negotiation*, notes that inefficient and undesirable outcomes have often resulted from 'the way in which negotiations have been structured and organized' (Boyer, 1999, p102).

The assumption, however, is a modest one. This book does not claim that organizational factors by themselves can account for the success or failure of a negotiation, or even that they are a consistently dominant factor. At the time of writing, US repudiation of the Kyoto Protocol and delayed Russian ratification were far higher on the list of forces shaping the climate negotiations than the qualities of the presiding officers, the use of informal groups or the timing of release of a negotiating text. Even the best-organized negotiation will fail if the political will to reach agreement is simply absent. This, however, does not mean that the organization of a negotiation process is unworthy of attention. The way a negotiation is organized is one of many, often competing, factors that together shape the negotiation process and direct it towards a particular outcome. The attraction of studying organizational factors lies in their openness to collective policy manipulation. While the extent of possible manipulation varies from case to case, short term organizational decisions can be taken that directly influence the course of negotiations, whereas other forces at play in a negotiation (such as distribution of power, domestic interests, the salience of an issue, broader geopolitics) rarely lend themselves so readily to such manipulation.

Global negotiations are also qualitatively different to other types of negotiations. The large number of participants, their economic, cultural, social, political and linguistic differences, and the historical baggage of international relations, all make for an inherently highly complex process. As well as complexity, global negotiations must also contend with the contradiction between the massive inequality of participants in terms of wealth and power, and the international norm of sovereign equality. These two defining characteristics of complexity and inequality raise enormous challenges for global negotiations, notably a tendency to inefficiency and strong competitiveness among negotiators, along with struggles over transparency and procedural equity. The way in which the process is organized is therefore far more important for global negotiations than for any other type. While two-party negotiations can run as a free-for-all, global negotiations simply could not function without procedural rules, institutional arrangements and organizational leaders.

This book is linked into broader, long standing debates over global governance, and global environmental governance in particular (e.g. see Rosenau and Czempiel, 1992; Young, 1997; Brack and Hyvarinen, 2002). Debates over global environmental governance span a range of proposals, including the potential for exploiting synergies across different regimes and even the possibility of establishing a new overarching Global Environment

Organization. Although the *negotiation process*, and how it could be improved, has remained on the sidelines of these debates, its rightful place is at their centre. In whatever way the current system of global environmental governance is eventually reformed (or not), it will remain founded on negotiation as the engine of intergovernmental cooperation.

The organization of the negotiation process is one of the lesser studied factors influencing the course of a multilateral negotiation, having enjoyed relatively little or only superficial attention in the academic literature compared with more popular lines of analysis, such as the impact of power structures, interests or knowledge (e.g. see Litfin, 1994; Rowlands, 1995; Zartman and Rubin, 2000a). This relative neglect can be attributed, at least in part, to the fact that academic researchers rarely have access behind the scenes of global negotiations where most decisions on organizational matters are made. In consequence, researchers tend not to be aware of the considerable effort that goes into organizational decision-making, of why certain options are implemented and others rejected, and of the implications of these decisions.

My objective in this book is to start to fill this gap in the understanding of the organization of global negotiations. In doing so, I draw heavily on my first hand experience of the organization of global negotiations gained while working for the UNFCCC secretariat during 1996–2001, as both a staff member and a consultant. I supplemented this first hand experience with two rounds of interviews, conducted in 1999/2000 and 2003/2004, with a range of participants in the climate change negotiations, including country delegates, non-governmental organizations (NGOs) and secretariat staff. Most of the 30 interviewees would only speak on condition of strict anonymity, so their names and affiliations are not given. In addition to these structured interviews, I have enjoyed, over the years, countless illuminating conversations with very diverse friends and colleagues in the climate change community. Much of the analysis in this book has been shaped and inspired by these private discussions.

The book explores the organization of the climate change negotiations from the first session of the Conference of the Parties (COP 1) to the UNFCCC in 1995 to COP 9 in 2003. The climate change negotiation process went through three distinct phases over this eight-year period:

- the *Kyoto Protocol negotiations* that led to the adoption of the Protocol at COP 3 in 1997
- the *post-Kyoto negotiations* on the details of the Kyoto Protocol and the implementation of the UNFCCC, which broke down at COP 6 in 2000, but finally culminated in the adoption of the political Bonn Agreements at COP 6 (part II) and the more technical Marrakesh Accords at COP 7 in 2001
- the *post-Marrakesh negotiations*, which continue to this day on more routine aspects of the development of the regime.

These phases make up a useful set of cases for the study of global negotiation processes, as they comprise both major successes (COP 3, COP 6 (part II), COP 7) and a failure (COP 6 part I), along with sessions that resulted in substantive agreement, but dissatisfaction over the negotiation process itself (COP 2, COP 4).

The book focuses on specific organizational elements, examining how these were manipulated and how they played out through the different phases of the climate change negotiations. Firstly, it looks at the roles of what might be called the *organizers* of the negotiation process:

- the presiding officers
- the bureau
- the secretariat.

Secondly, it analyses the organizational rules governing the negotiations, including:

- rules for the conduct of business
- rules for decision-making.

The book then focuses on three critical dimensions of the negotiation process – place, time and the written word, that is:

- the negotiating arenas and complementary (non-negotiating) forums (such as workshops)
- the negotiating texts
- the timing of the negotiations.

Finally, it explores the channels for eliciting, receiving and processing input from two important, but very different, sets of participants:

- ministers
- NGOs.

Two opening chapters frame the analysis. The first explores in more detail the challenges faced by global negotiations and the nature of the organizational elements (Chapter 2). The second opening chapter introduces the climate change issue, examining the specific challenges faced by the climate change negotiations and providing an overview of the various negotiating phases to date (Chapter 3). The book ends by drawing some general conclusions on the organization of the climate change negotiations, including 12 key lessons that could be applied to global negotiations more generally.

The Organization of Global Negotiations

Having started in a very civilized fashion with songs about the future from children's choirs...the meeting...finished at four o'clock in the morning, one day late, with most of the delegates having abandoned their chairs...to gather on the front podium and shout at each other (Brenton, 1994, p183).

Introduction

This chapter sets out the analytical framework for this book and introduces key concepts. It begins by discussing negotiations in generic terms, before turning specifically to global negotiations, and analysing the challenges that these face. The chapter then examines how regimes and negotiations relate to each other, before exploring more fully the concept of the organization of the negotiation process.

The fundamentals of negotiations

A negotiation can be understood as 'a process of mutual persuasion and adjustment which aims at combining non-identical actor preferences into a single joint decision' (Rittberger, 1983, p170). Understood in this way, negotiations are a ubiquitous mechanism for decision-making at every level of social interaction and among a range of actors. Family members negotiate on their holiday destination, traders barter over goods, trade unions negotiate with employers over pay deals, governments launch negotiations on cross-border concerns.

For any negotiation to take place, three key conditions must be fulfilled. Firstly, two or more actors must be in a situation of *interdependence*, that is, they must share an area of common interest where the actions of one will affect the other(s). Secondly, their interdependence must be characterized by *discord*, that is, with the actors preferring different courses of action. These two conditions are fundamental; 'without common interest, there is nothing to negotiate for, without conflict, nothing to negotiate about' (Iklé, 1964, p2). Thirdly, the actors involved must (implicitly or explicitly) *eschew other means* of resolving their case of discordant interdependence, notably, the use of overt force or having recourse to an independent adjudicator.

The negotiation process will typically pass through a number of stages. Firstly, participants will need to agree the *agenda*, or *mandate*, for the negotiation, setting out, for example, the issues to be covered, the deadline, and the main forum for the talks. Agreeing these basic conditions for the negotiation may in itself be contentious. The negotiation process proper will often start with a period of *exploration* of the issues on the agenda, which may involve gathering data and information for collective analysis by the negotiators, or calling for independent technical input.

A key stage in the negotiation will be when participants make formal *proposals*, expressing divergent preferences, or positions, on at least some of the issues under negotiation. The divergent positions will tend to reflect a mix of both tangible perceived interests and intangible values and principles, of which the latter are likely to be particularly stable and less amenable to change. The tabling of proposals will often mark a transition from exploration to *bargaining*, after which negotiators explicitly engage with one another's positions to devise solutions that can bring them to agreement, through mutual compromises, trade-offs, linkages, packaging, side-payments, adding and subtracting issues and so on.[1] Bargaining will be relatively low key at first, focusing on more peripheral, less contentious issues, becoming more intense over time. The final stage in a negotiation is *deal-making*, where negotiators will start to reach agreement on the issues on the table, starting with the more straightforward ones. Negotiations traditionally end in a dramatic finale, where the more difficult questions are finally tackled through intensive bargaining. It can often seem as if almost all the real work is done in the finale. The negotiation will end when negotiators conclude, often in the face of a deadline, that they have adjusted their positions, and others have adjusted theirs, as much as they can. If, at this stage, the compromise solution on the table is perceived as preferable to what the negotiators could obtain without an accord, then substantive agreement is likely to be reached.

Global intergovernmental negotiations: Challenges

While all negotiations share the same fundamentals, specific negotiation types merit separate study. These are usually differentiated by the *type* of negotiating parties (e.g. individuals, labour organizations, governments) and their *number*. The particular negotiation type upon which we focus is a *global intergovernmental negotiation*, that is, where the main negotiating parties are sovereign states and the negotiation involves many such states.

State governments are intricate negotiating parties. As individual negotiators, government representatives in an intergovernmental negotiation are accountable to their domestic legislatures. This is not unusual; most negotiators, at any level, negotiate on behalf of a constituency, for example, a trade union. However, in negotiations among states, the relationship between the individual negotiator and his/her constituency is particularly complex.

States are not monolithic, and different government departments, as well as individuals within departments, are likely to have varying views over the national positions to be taken to the intergovernmental arena, and then over whether, when and how to adjust those positions as negotiations progress. The position of a particular state will also be influenced by domestic political considerations, which may be only tangentially related to the issue under negotiation. Differences between government departments within a state are likely to be reflected in (usually sharper) differences among NGOs who will lobby the state – often in different directions – to take a particular stance at the negotiating table. In short, government negotiators must reach agreement at the domestic level to define their preferences before they can negotiate in the multilateral arena. Putnam (1988, p434) has famously conceptualized this as a 'two level game':

> ... at the national level, domestic groups pursue their interests by pressuring the government to adopt favourable policies, and politicians seek power by constructing coalitions among those groups. At the international level, national governments seek to maximize their own ability to satisfy domestic pressures, while minimizing the adverse consequences of foreign developments.

Another distinguishing feature of intergovernmental negotiations is that they are inherently repetitive – repeated games, in the language of game theory. That is, 'each government knows that it will always be engaged in negotiation' (Iklé, 1964, p90) with other governments on a whole variety of topics within the broad intergovernmental arena for the foreseeable future. The dynamics of the specific negotiation at hand, therefore, will be influenced by, and will in turn impact on, the wider geopolitical network of relationships and interactions between states beyond the particular problem under negotiation.

Turning to the number of negotiators, the negotiation literature makes a major distinction between two-party (bilateral) and more-than-two-party (multilateral) negotiations (e.g. see Raiffa, 1982; also Zartman, 1994). Among multilateral negotiations, Midgaard and Underdal (1977) identify three different types: up to 7 parties (small), 7–20 parties (intermediate) and 20+ parties (large). If Midgaard and Underdal had been writing 20 years later when such negotiations had become much more common, they would probably have added a fourth category, that is, intergovernmental negotiations whose potential scope includes all or almost all existing state governments. This type of multilateral negotiation, which is also known in the literature as 'conference diplomacy' (e.g. Rittberger, 1983; Kaufmann, 1989, 1996), is termed here a *global* negotiation (Kremenyuk and Lang, 1993).

Two linked trends have seen the number and profile of global negotiations rise. Firstly, the emergence in the 1980s of environmental problems with a global reach – such as stratospheric ozone depletion, biodiversity loss and, of course, climate change – triggered the launch of negotiations open to all states (often under UN auspices) to agree legally binding instruments to tackle them. Secondly, during the 1990s, the UN convened a series of major global conferences on far-reaching topics of global interest. These included, for example, summits on population (Cairo, 1994), women (Beijing, 1995), social development (Copenhagen, 1995), cities (Istanbul, 1996) and, of course, the UN Conference on Environment and Development – the Earth Summit – in Rio de Janeiro in 1992. These two trends reflected as much a belief in the promise of multilateralism as a basis for rebuilding international relations in the wake of the collapse of the post-war East/West divide as the truly global nature, or otherwise, of the problems under negotiation.

Complexity and inequality

A key defining characteristic of, and challenge for, global negotiations is their *complexity* (Winham, 1977; Zartman, 1994; Hampson and Hart, 1995). This complexity takes many forms. The large number of negotiating parties – that is, national governments – which are represented means a large potential range of different positions and proposals to present, discuss and reconcile. While it is, of course, theoretically possible that even a large number of negotiating parties might share similar views, the fact that a negotiation is taking place at all implies disagreement, while the heterogeneity of states, in terms of their political, social, economic and geographical circumstances, increases the potential diversity of opinion.

An important element of complexity is that, unlike a bilateral negotiation where negotiators are clearly aware of who their opponent is and to whom they must target their negotiating strategies, there is no such inherent structure to a multilateral negotiation. Negotiators must seek to induce many counterparts with varying points of view to adjust their preferences, decide to whom they should concede what, and assess what the knock-on effects that concession might have on the preferences of the other counterparts, as well as on support from their domestic constituencies.

To add to the complexity, government delegations to a global negotiation usually consist of more than one person (often very many more), who may themselves have differing personal views and negotiating styles. Although global negotiations take place among states, the negotiators that represent those states are still human beings, and the personalities of those human beings, along with the interactions among them, can be as important as it can in a bilateral negotiation between employee and boss. According to Lang, 'personalities determine the course of negotiations. These personalities are not abstract beings: They suffer from fatigue; they may lose their temper; they may feel frustrated; they take pride in accomplishing a specific, arduous task' (Lang, 1991, p389).

The sheer number of individuals present in a global negotiation, often numbering into the thousands, in turn creates massive organizational challenges in simply making sure that everyone is in the right place at the right time, working from the same agreed agenda in a common language with the necessary documentation, and with the opportunity to express a view and be heard. Interaction among those individuals takes place among a wide variety of cultures, language groups and negotiating styles, creating obstacles to effective and constructive communication.[2] As Kaufmann notes, 'the mere multiplication of the number of recipients of messages renders the establishment of effective communication much more difficult in conference diplomacy than in traditional bilateral diplomacy' (Kaufmann, 1989, p173). The complexity of the social context in which negotiators are operating means they must develop 'short cuts' to make sense of that context. Negotiators will therefore tend to stereotype each other, interpreting messages they receive to bolster their existing (often negative) perceptions rather than exploring the intended (often more positive) meaning behind them. Any observer of a global negotiation will pay testimony to the perceptual distortions that often prevail, especially between groups that traditionally distrust each other, such as the developing country Group of 77 (G-77) and the industrialized states.

A common tendency within global negotiations to try to cope with the complexity of many parties is the formation of coalitions. These may be more or less formally articulated, ranging from coalitions who almost always negotiate together and speak with a single voice, to looser groups confining themselves to information sharing. Coalitions can help to reduce somewhat the complexity of a global negotiation by cutting down the number of proposals or requests to speak. They introduce their own dynamics, however, which may complicate and slow down the negotiation process. In basic terms, coalitions introduce a third level to the two level game, so that negotiating parties must reach agreement with their coalition, as well as with their domestic constituencies, before coming to the main negotiating table. Some coalitions, especially large ones or those that negotiate in close unison, may find it difficult to reach agreement among themselves, or respond promptly to developments in the negotiations. Coalition positions also tend towards the lowest common denominator, and may entrench wider political cleavages, such as the historical divide between developing and industrialized countries, as they take refuge in ideology or shared political history to cement their group loyalty.

Almost by definition, a large number of negotiating parties – even if they are organized into coalitions – implies a large number of *issues* on the table, as governments put forward different proposals. According to Homans' Maxim,[3] a large number of issues could, in theory, generate greater potential for achieving an acceptable outcome with joint gains for all. This potential, however, can often fail to be realized through sheer inability to manage the resulting complexity.

The complexity of global negotiations raises their transaction costs, that is, the costs incurred to reach agreement in physical and financial resources, human effort and time. This tendency to high transaction costs – in other words, inefficiency – of global negotiations is well known, and is reflected in the stereotypical view that 'modern intergovernmental conferences of the UN General Assembly ... are a waste of time, energy and money' (Kaufmann, 1989, p1). The tendency to inefficiency is a major challenge for global negotiations.

Another important feature of global negotiations is the *inequality* of their participants. Unlike complexity, inequality is not inherent to a negotiation with many parties, although it is very common. In a large negotiation, it is unlikely that all parties will possess even roughly equal power, whether defined in economic or political terms. In a global intergovernmental negotiation, involving all the world's states, inequality is pervasive, whatever the issue under negotiation. The fundamental differences in economic wealth and resources between states are inescapable, and affect the substantive leverage one country has over another. Smaller, poorer states may find unexpected sources of power (e.g. moral authority among the small island developing states in the climate change negotiations), but this rarely makes up for their fundamental lack of leverage in broader geopolitical and economic terms. Similarly, a one person delegation will, unless that person is extremely skilled (which some are), find it much harder to exert influence in a negotiation than a delegation of 20 persons. Inequality, of course, has a history, and this can contribute to mistrust among negotiating parties. An important dimension of inequality is its contrast with the *formal equality* of nation states as sovereign entities within most international institutions. While sovereign states are all equal in international law, raising expectations that they should be treated as such, in practice the differences among them in terms of their political and economic power and capacities render it difficult to realize such formal equality.

Procedural inequity

Substantive power inequalities tend to be reflected in *procedural inequity*, that is, the unequal capacity of parties to participate effectively in the negotiations. In most global negotiations, parties enjoy *formal* procedural equity, expressed through a one-state-one-vote system. In practice, however, the capacity of parties to participate in the negotiations, and therefore their *practical* procedural equity, varies tremendously. A crude measure of this is delegation size. A large delegation means that more individuals are available to cover the many issues under negotiation, to build relationships with other parties and engage in behind the scenes talks, as well as to analyse the implications of proposals and develop well thought out positions based on them. Smaller delegations find it much more difficult to negotiate effectively in this way, as they are spread much more thinly. The gap between formal procedural equality and practical procedural inequality poses challenges to the success of global negotiations. The legitimacy of the process and the acceptability of the outcome may be compromised if the negotiations are not perceived as fair, while 'countries that do not perceive a process to be fair have great power to obstruct it, ensuring that negotiations make little progress' (IPCC, 1996, p117).

Transparency concerns

Practical procedural inequity, coupled with complexity, has a tendency to erode *transparency* in global negotiations. Faced with a large number of negotiating parties with many divergent views and proposals on the table, the more powerful countries, the big players, will often be tempted simply to shut out their less powerful negotiating adversaries to focus on discussing what they consider to be the key issues and options, and then 'to confront the other partners with a finished agreement which it will be difficult to oppose' (Iklé, 1964, p135). The fact that parties are typically negotiating from a position of formal procedural equity expressed through sovereign equality, however, triggers a demand for transparency, that is, for all negotiating parties to be involved in, or at least to be kept informed of, developments in the negotiations, and for all issues to be presented for debate in a fully open forum before they are decided upon. As Freymond put it, 'one of today's challenges' is therefore 'how to reconcile the relatively greater efficiency of a negotiation limited to selected parties with the "democratic" right of sovereign states to be associated with decisions likely to affect them' (Freymond, 1991, p131). As with procedural equity, maintaining as much transparency as possible is important to upholding the legitimacy and acceptability of the process. Negotiators are more likely to feel a sense of ownership of an eventual agreement negotiated in a transparent manner, increasing the likelihood of its ratification and implementation back home.

A second dimension to transparency – transparency relative to public scrutiny both inside and outside the conference venue – is a particular challenge for global *environmental* negotiations, given the high degree of public interest in, and concern over, these issues. Indeed, global environmental negotiations do tend to be more open to the public than negotiations on more traditional economic or security issues, or bilateral negotiations. 'Public' in this regard includes NGOs, such as environmental and business groups and other advocacy or academic organizations, as well as the media, through which the interested public outside the conference venue is informed of proceedings. Meeting demands for transparency can bring benefits to a global negotiation, potentially increasing pressure on negotiators to reach agreement and promoting eventual implementation. In addition, participation by environmental NGOs can provide a 'voice for the environment', thus promoting an environmentally stronger agreement. Public transparency is also generally viewed as a good in itself, holding representatives of states accountable to their domestic constituencies.

There is a downside to transparency, however. Public scrutiny can make it more difficult for negotiators to climb down from their positions and explore proposals in a creative manner, while making provisions for NGO participation can also aggravate the complexity and inefficiency of a negotiation. This is especially the case as NGOs are themselves heterogeneous, and often seek to influence government negotiators in sharply contrasting ways. Some may even seek to obstruct the process. This may generate what Wettestad calls 'a classic dilemma ... related to the conflicting general concerns for openness and legitimacy versus decision-making effectiveness' (Wettestad, 1999, p21). Indeed, the debate over the extent to which negotiations should be secret or open is a longstanding one in the diplomatic arena (e.g. see Nicolson, 1939, 1963; Iklé, 1964).

Competitiveness

Another challenge arising from the complexity and inequality that characterize global negotiations is a *tendency to competitiveness*. In cooperative behaviour, negotiators engage in joint problem solving to collaboratively explore solutions that achieve an integrative, or win–win, solution: in other words, joint gains for all. In competitive behaviour, however, negotiators engage in individualistic or positional bargaining, seeking to defend positions and extract maximum gains from one another through a more confrontational, win–lose approach. Competition and cooperation are not absolutes, of course, and varying degrees of both are possible.

There is an assumption in the literature that a more cooperative, problem solving approach is beneficial to a negotiation, being capable of reconciling more fully the preferences of all parties, as well as generating better relations among them (e.g. see Fisher et al, 1992; Underdal, 1991b). Competitive behaviour, where a negotiation becomes a battlefield and other negotiators are viewed as adversaries, is likely to exacerbate tendencies among negotiators to want to save face or score points, reducing their willingness to consider the proposals of others, concede or accept compromises. The potential disadvantages of competitive behaviour are particularly high in global negotiations, where inherent complexity raises the transaction costs of such an approach relative to cooperation. As Fisher et al note, 'the more people involved in a negotiation, the more serious the drawbacks ... If some 150 countries are negotiating, as in various United Nations conferences, positional [competitive] bargaining is next to impossible' (Fisher et al, 1992, p7). In order to reach agreement through competitive behaviour, individual negotiators or coalitions will need to extract concessions from their many counterparts, while each of their counterparts are trying to do the same thing with the others. This makes for a chaotic and time-consuming process, where the interests of all parties are unlikely to be integrated.

However, despite the theoretical potential for cooperation and the costs of competitive behaviour, it is the latter approach that tends to dominate in global negotiations, exacerbated by the misunderstandings and confused messages that can prevail in a situation of such complexity. According to Martinez and Susskind:

> Bargaining in the international arena is intrinsically positional [competitive]: negotiators...arrive...with carefully crafted marching orders – from which they are not supposed to deviate. Their stated 'positions' are, for the most part, not open to revision without consultation with...domestic ministries. Even though [negotiators] recognize that the invention of additional 'packages' might well produce better results for all sides, they are allowed precious little leeway at the negotiating table. The risk that something offered in an informal exploratory exchange might be misinterpreted as a commitment or misused by others is too great (Martinez and Susskind, 2000, p571).

At the extreme end of competitiveness, there are cases where parties negotiate with the implicit aim to obstruct agreement, what Wallihan (1998) calls 'avoidance bargaining'. Such parties tend to exert a disproportionate influence on the negotiation. As Fisher et al note, 'the relative negotiating power of two parties depends primarily upon how attractive to each is the option of not reaching agreement' (Fisher et al, 1992, p106). Given that such obstructionist parties actively seek non-agreement, their bargaining leverage in the negotiation can be high.

Regimes and negotiations

Global negotiations are often closely associated with the formation and development of *regimes*, defined as sets of both formal and informal rules, institutions and procedures aimed at governing action in a particular issue area, usually based on a founding treaty. Regime formation has become a popular response to emerging global environmental problems over the past two decades, with the establishment of regimes to address, for example, stratospheric ozone depletion, climate change, biodiversity loss, desertification and persistent organic pollutants. Such regimes have been constructed through negotiations among states, usually within a wider institutional setting, typically the UN or one of its agencies or specialized bodies.

Once established, regimes are rarely static. On the contrary, most regimes are continuously evolving through decisions by their state parties. Such evolution may include, for example, elaborating new legal texts based on the parent treaty such as protocols or amendments, a process that could be termed *regime strengthening*. The more routine evolution of regimes, which might be characterized as *regime development*, involves less momentous, but still significant, decisions, such as the preparation of guidelines for national reporting or the establishment of new expert bodies. The regime strengthening and development processes again take place through the mechanism of negotiation, based on the rules and institutional arrangements of the regime, thus connecting regimes and negotiations in an iterative relationship. Another form of negotiation that often accompanies regime development or precedes regime strengthening is *regime review*, where the negotiating parties exercise oversight over how the regime's rules are working and being implemented. This can involve both routine review – for example, considering the periodic national reports submitted by parties – or a more wide ranging review of the commitments of parties or the functioning of the regime, whose results are expected to lead to a new round of regime strengthening.

In many cases, a *continuous negotiation process* emerges within a regime. Indeed, an important function performed by regimes is precisely to provide an efficient framework for negotiations so that, as Keohane puts it, regimes become 'devices to make agreements possible' (Keohane, 1989, p111). Regimes and negotiations are therefore intimately related. Both are concerned with cooperation among states to address problems of interdependence but, while regimes draw attention predominantly to the *structure* of cooperation, negotiations are mostly concerned with the *process* that unfolds based on that structure.

Negotiations within regimes thus have a special character. Like all intergovernmental negotiations, they are repetitive, but more intensively so, often involving not only the same governments, but also the same individuals who regularly work together for many weeks, year after year. The negotiation process within the regime typically gives rise to its own set of informal practices and procedures, even its own culture. While such intensively repeated games can provide important opportunities for learning and therefore improving ways of negotiating, the flipside is the danger of ossification; the negotiation process gets stuck in old ways of thinking and doing that drag down substantive progress. The inherent complexity of global negotiations, combined with the inequality of participants, can block out new thinking and contribute to such entrenchment.

The organization of global negotiations

The focus of this book is on the organization of global negotiations, and how organizational factors can help (or not) to overcome the challenges faced by global negotiations.

Before proceeding further, we need to pin down exactly what is meant by 'organization' in this context. Organizing a global negotiation involves managing a range of organizational elements. Organizational elements that tend to be of particular importance to global negotiations, and which are covered in this book, are as follows:

- Rules for the conduct of business and decision-making
- The use of different arenas for negotiation and discussion (e.g. big/small, open/closed, inclusive/limited participation)
- The timing of the negotiations
- The use of negotiating texts
- Rules for high-level participation (e.g. by ministers)
- Rules for the participation of stakeholders (e.g. environmental NGOs, businesses).

This list of elements is admittedly not comprehensive. The organization of the negotiation process is not an easy phenomenon to delimit (especially as there is little help in the literature on how to do this), and other organizational elements could also come into play, such as the geographical venue of a negotiation, the financing of the regime, the way parties organize themselves into coalitions and the composition of delegations. This book, however, focuses on what Young calls 'decision variables', that is, 'factors which are subject to conscious control or manipulation on the part of those responsible for designing and managing international regimes' (Young, 1994, p152). It is generally not possible, for example, to make a collective decision regarding the coalitions that parties should form or the composition of their delegations, as this choice is based on individual, sovereign choice. If the aim is to improve the organization of negotiation processes, then it makes sense to focus on those factors that can be subject to deliberate manipulation.

The governance of organizational elements

The management of the above-mentioned organizational elements does not take place in a vacuum. Instead, it is governed by a hierarchy of formal and informal institutions, and procedural rules.

Founding texts

The founding treaty of the regime will itself establish a set of institutions within which subsequent negotiations take place, usually a Conference or Meeting of the Parties, with one or two subsidiary bodies. The founding treaty may also include a limited set of procedural rules, such as voting majorities for decision-making in specific circumstances, or basic rules for admitting NGOs. These are very stable organizational elements, which, inscribed as they are in a legal treaty, tend to be well-respected and rarely subject to change.

Rules of procedure

One of the first acts of the newly established Conference (or Meeting) of the Parties[4] to a regime will be to agree formal rules of procedure for itself and its subsidiary bodies. These will set out in more detail rules for the conduct of business and decision-making, the role of presiding officers, the secretariat and bureau, and also procedures for the participation of observers. Once agreed, these rules are likely to remain stable, although they may be subject to formal amendment.

Supplementary decisions

The rules of procedure may be supplemented by additional decisions taken by the COP or subsidiary bodies on organizational issues. These may be ad hoc decisions that apply only to a specific negotiating session (e.g. the format for ministerial participation at the next COP) or more general decisions that apply to all subsequent sessions (e.g. access rules for closed negotiations).

Informal practices

The formal procedural rules are rarely comprehensive, and are often complemented by informal practices that develop spontaneously over time. Informal practices may emerge to interpret ambiguities and fill in gaps in the formal rules, or they may arise as actors improvise in response to changing circumstances. Different informal practices will be more or less established. Some may become as deeply entrenched as formal rules, while others may be used at just one or two negotiating sessions.

Wider institutional setting

The formal and informal procedural rules at play in the wider institutional setting – typically the UN General Assembly or other UN body or agency – will exert an important influence, not only on the shaping of the above formal institutions, rules and procedures, but also on the interpretation and improvisation that surrounds them.

Who organizes?

Discussing the organization of a negotiation process begs the question 'who organizes?' The organization of the negotiation process is indeed a conscious act. Formal rules – whether inscribed in a treaty, rules of procedure or supplementary decisions – are the products of decisions taken by the negotiating parties, while the development and application of informal practices is also subject to the consent of the parties. However, the key individuals involved in the day-to-day organization of a negotiation are typically its *presiding officers, bureau and secretariat*, to whom the negotiating parties delegate organizational decision-making. The basic roles, functions and scope of authority of these organizers are themselves established in the founding texts and rules of procedure of the regime, sometimes elaborated on in supplementary decisions, and put into play through improvisation, interpretation and informal practices.

Depending on their skills and attributes, and the needs of the negotiations, the organizers can assume very important roles in supplying *process-oriented leadership*, or *organizational energy*, to the process. Process-oriented leadership (Wettestad, 1999) refers to leadership seeking to promote the broad success of a negotiation. It is therefore distinct from substantive leadership aimed at furthering a particular outcome desired by the leader, which would more likely be exercised by a negotiating government. Exercising process-oriented leadership can be a key task of presiding officers and, to a lesser extent, the bureau, especially in the context of a vacuum in (more often substantive) leadership on the part of any party.

A related, more expansive means of conceptualizing process-oriented leadership is in terms of the provision of *organizational energy* (see Underdal, 2002). Leadership requires 'followership' (Rubin, 1991); that is, it is a self-conscious act of taking control whereby followers know and consent to being led. The supply of organizational energy, however, can be more subtle and diffuse, being injected by actors seeking to promote the negotiating goal but without assuming, and sometimes deliberately eschewing, the mantle of explicit leadership. This encapsulates well the role of the secretariat, and also applies to the bureau, whose lower profile work is typically practised behind closed doors, rather than through the public acts of leadership characteristic of presiding officers. There is, of course, a fine line between energy and leadership. While it is interesting analytically to distinguish between these, the key point is that both involve providing process-oriented and organizational input to the negotiation process that is aimed at advancing its success.

The organization of the negotiation process is certainly not a neutral or purely technical act. How a negotiation is organized can influence not only the effectiveness of the process, but also the relative advantages of individual negotiating parties. This is reflected in the notion of 'organizationally dependent capabilities' (Keohane and Nye, 1977; Keohane, 1993), which draws attention to the fact that the particular structure of regimes and, by extension, the organization of the negotiation process, can bring relative advantage to some actors and disadvantage to others, for example, through its decision-making procedures or use of closed versus open forums. Certain aspects of the organization of the negotiation process can thus become hotly debated, as parties promote organizational forms in strategic ways to advance their interests or use organizational issues as an outlet for national positions and grievances.

Summary and concluding remarks

In summary, the organization of the negotiation process consists of several layers of more or less formal sets of institutional arrangements, procedural rules and informal practices, which are actively organized by parties, but especially by the organizers of the negotiation process – the presiding officers, bureau and secretariat. This book explores how organizational factors have been used in different ways over the history of the climate change regime to try to overcome the challenges of global negotiations. These challenges, as discussed above, are centred on complexity and inequality, generating dilemmas relating to procedural equity, transparency and tendency to competitive behaviour.

The Challenges of the Climate Change Negotiations

It was the mother of all negotiations[1]

Introduction

This chapter provides an introduction to the climate change negotiations, drawing on the framework for understanding global negotiations set out in Chapter 2. It begins by outlining the nature of the climate change problem and the characteristics that render it particularly difficult to address. The chapter then turns to the climate change negotiations themselves. Although the focus of this book is on the period 1995 to 2001, incorporating the Kyoto Protocol negotiations and the post-Kyoto process that followed, the chapter includes a comprehensive overview of the negotiation process from its inception to the present day. We then explore the challenges posed by the climate change negotiations, revolving around the themes of complexity and inequality, and how these challenges have evolved over time. These are the challenges that the organizers of the negotiation process must deal with in order to promote successful outcomes.

The climate change problem

Climate change is commonly viewed as a uniquely 'malign' problem (see Wettestad, 1999; Miles et al, 2002), presenting 'the decision maker with a set of formidable complications' (IPCC, 1996, p7). While there is a tendency for all negotiators to see 'their' issue as the most difficult in the international arena, climate change does have a good case for this distinction.[2]

Firstly, climate change is the most global of the global environmental problems. Greenhouse gas (GHG) emissions are well mixed in the atmosphere, which means that there is no relation between the geographical location of emissions and their eventual effects. All countries, therefore, are potential victims of climate change, although their vulnerability will differ considerably depending on their geographical circumstances and economic wealth (Yamin and Depledge, 2004). Moreover, the global pervasiveness of the main drivers of climate change – fuel combustion and land-use change – means that all countries, all regions, and all individuals have the potential to contribute to the problem, even though, as discussed below, they do so to varying extents. This means that, to develop an effective response to climate change, the negotiations do need to involve all the world's states, or at least a critical mass of those most responsible for, and potentially vulnerable to, the problem.

Uncertainty, both scientific and economic, is a pervasive and very difficult feature of decision-making on climate change. From a scientific perspective, the climate system is immensely complex, naturally variable and chaotic, while multiple feedback mechanisms, both positive and negative, are at work. The concentration of GHGs in the atmosphere is only one factor among many (e.g. ocean currents and solar radiation) determining global temperatures, so that it is difficult to identify the signal of human interference against the background noise of natural climate variability. Estimating how average global temperature trends will translate to changes in local weather patterns, such as rainfall and the incidence of weather extremes, is particularly complex. Uncertainty is thus compounded by indeterminacy, so that causally linking a potential climate catastrophe to carbon emissions from, for example, the family car is a difficult task.

Uncertainty also reigns over how to respond to climate change. The predicted costs of achieving various emission targets, as well as the costs of climate change impacts, depend on the assumptions and methodologies of the economic models used, and therefore vary greatly. According to the Intergovernmental Panel on Climate Change (IPCC), 'estimates have spanned such a wide range that they have been of limited value to policy making' (IPCC, 1996, p303). In addition to the well-known problem of differing assumptions and methodologies, the sheer nature of climate change poses real challenges for conventional economic analyses, including its long time-lags, equity dimensions, and potential for catastrophic change, as described below.[3]

The pervasiveness of fossil fuel combustion and land-use change, the two major drivers of climate change, means that the problem is implicated in almost every human activity. GHGs are emitted from a myriad small and large sources, from power plants to private cars, paddy fields to gas cookers, aluminium smelters to land-fill sites.[4] There is, therefore, no single identifiable cause of climate change and no single salient solution. The ubiquity of the causes of climate change, in turn, means that almost any group within civil society could be considered a stakeholder, thereby generating great public interest in the topic, even by the high standards of global environmental issues. The economic implications of climate change have, in particular, motivated a strong presence on the part of business and industry groups. As Barrett and Chambers comment:

> international climate change negotiations have involved a greater number of actors as well as many more kinds of actors than any of the numerous sets of multilateral negotiations that have occurred in the post-war era (Barrett and Chambers, 1998, p15).

Climate change has a very long time horizon due to the inertia of the climate system and the long lifetime of some GHGs.[5] This raises issues of intergenerational equity, as it is future generations that will suffer the worst consequences of past and present emissions.[6] Conversely, mitigation of climate change calls on present generations to incur short-term costs whose long-term benefits will only be enjoyed by future generations, creating disincentives to action in the present. The long time-lags involved have also created a sense of unreality about climate change, which has rendered it more difficult to persuade policy-makers and the public to take the problems seriously.

Climate change is fundamentally a development issue, as indeed are all global environmental problems. The development dimension of climate change is, however, particularly acute, given that fuel combustion and land-use change are intimately entwined with the process of modernization. Climate change thus challenges the mode of development followed since the industrial revolution. Although views differ over the costs of climate change mitigation, the changes required to prevailing economic and social structures and dynamics are indisputably great, and the political stakes – in both addressing climate change and not doing so – are therefore high.

Another aspect of the development dimension of climate change is its inherent unfairness, which raises further questions of equity. The smallest contributors to climate change, generally the poorer developing countries, are also the most vulnerable to climate change impacts, as they tend to be more reliant on agriculture, have weaker infrastructure and lack the resources to take adaptation measures, such as building sea defences. Many are also located in geographically vulnerable regions (e.g. low lying coastal zones), or are already suffering from environmental stress (e.g. from desertification or overpopulation). By contrast, the larger emitters, mostly the richer industrialized countries, are typically less vulnerable, possessing modern economies that do not rely so much on the vagaries of the weather, more resilient infrastructure and more abundant resources to adapt. As Grubb provocatively puts it, 'greenhouse gas emissions involve the rich imposing risks upon the poorer and more vulnerable' (Grubb, 1995, p467).

Finally, an important characteristic of climate change is the potential for irreversible and catastrophic change. Irreversibility is a function of the long time-lags involved, which mean that climate change could not be reversed in human time spans, while concern over catastrophic change is based on the possibility that rising atmospheric concentrations of GHGs could force the climate system into a different state in the relative short-term. Rising average global temperatures could, for example, force the Gulf Stream to change course or shut down altogether, thus leading to dramatic cooling in northern Europe, or large ice masses, notably the West Antarctic ice sheet, could collapse, generating massive sea level rise. The 2001 IPCC Assessment suggests that, while very unlikely this century, rising emissions over the next few decades could set these catastrophic events in motion. This potential for catastrophic change, whose impact would be immense but whose likelihood is uncertain, is particularly difficult for policy makers to integrate into their decision-making.

The climate change negotiations and the climate change regime

Despite its uniquely malign characteristics, governments were able to agree a regime, and embark on a continuous negotiation process centred on that regime, to address the problem of climate change.[7] The nature and structure of the climate change regime, and the phases of the climate change negotiation process to date, are discussed below. The focus of this book is on the period since the entry into force of the UNFCCC and the first Conference of the Parties (COP 1) in 1995, that is, comprising the Kyoto Protocol negotiations, the post-Kyoto negotiations, and latterly the post-Marrakesh negotiations (see summary in Table 3.1). For the sake of completeness, however, and to provide the necessary contextual and chronological background, the overview below also discusses the earlier phases of regime building, development and review up to COP 1.

Regime building. The foundation years: 1990–1992

In December 1990, the United Nations General Assembly (UNGA) launched negotiations on a framework convention on climate change to provide the foundation for a global climate change regime. These regime-building negotiations lasted nearly 18 months, culminating in the adoption of the UNFCCC on 9 May 1992. Key elements of the UNFCCC are summarized in Box 3.1 below.[8]

Box 3.1 Key elements of the UN Framework Convention on Climate Change

- defines an ultimate objective and principles.
- divides countries into:
  - Annex I (OECD countries and economies in transition - (EITs))
  - Annex II (OECD countries only); and
  - Non-Annex I (mostly developing countries).
- all parties:[1] general commitments, including obligations to prepare national communications.
- Annex I parties: specific 'aim' to return emissions to 1990 levels by 2000.
- Annex II parties: must provide financial assistance to developing countries, and also promote technology transfer, including to EITs.
- provisions for regular review of the Convention, including a review process to assess the 'adequacy' of Annex I party commitments at COP 1 and thereafter.

Source: Adapted from Yamin and Depledge (2004)
Note: [1] Countries that have ratified the Convention are known as 'Parties'

These initial regime-building negotiations were conducted under the auspices of the UNGA, and were therefore organized based on that body's formal procedural rules and informal practices. Drawing on the UNGA in this way

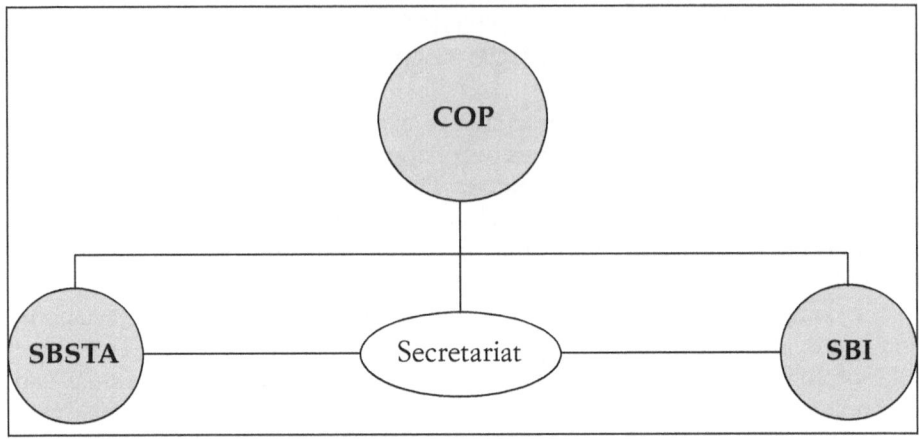

Figure 3.1 *The basic institutional structure of the climate change regime*

subsequently influenced the establishment and evolution of the climate change regime's own procedural rules and practices. These early regime-building negotiations also began to throw light on the political dynamics and key issues that would shape the continuous climate change negotiation process to this day. Notwithstanding the controversies and divisions that characterized them, the early days of the climate change negotiations took place against a background of considerable attention and optimism surrounding the commitment and ability of the international community to tackle global environmental problems. This came to a head at the June 1992 Earth Summit, where the UNFCCC was signed by 154 countries (and the European Community), including the United States. The UNFCCC thus became one of the three Rio Conventions, including the 1992 Convention on Biological Diversity that was also signed at the Earth Summit, and the 1994 UN Convention to Combat Desertification whose negotiation was launched in Rio de Janeiro.

As well as defining substantive commitments, the UNFCCC also established the basic institutional structure of the climate change regime, including: a supreme decision-making body – the COP; two subsidiary bodies that provide advice to the COP – the Subsidiary Body for Scientific and Technological Advice (SBSTA) and the Subsidiary Body for Implementation (SBI); and a secretariat, serving both the COP and the subsidiary bodies. This basic institutional structure, depicted in Figure 3.1 above, has provided the setting for subsequent negotiations within the climate change regime that continue to this day.[9]

Regime development and review. The prompt start negotiations and review of adequacy: 1992–1995

This first phase of regime development unfolded pending the entry into force of the Convention and the holding of COP 1, which was mandated to take place within a year of entry into force. These regime development negotiations, characterized as a 'prompt start' to the implementation of the Convention, focused on drafting decisions that would enable the Convention to be put into practice. This included, for example, drafting guidelines for Annex I parties to follow when preparing their first national communications (in other words, national reports) and a process for reviewing those reports; launching a pilot phase of activities implemented jointly (based on Convention provisions on the controversial concept of joint implementation – see below); setting out arrangements for cooperation with the Global Environment Facility (GEF), as the operator of the Convention's financial mechanism; detailing the respective roles of the two subsidiary bodies; and trying (in vain – see below) to agree rules of procedure. This prompt start regime development phase culminated with the entry into force of the Convention in March 1994, and the adoption of a set of decisions at COP 1 in April 1995.

Although regime development generated the greatest volume of work for this phase, the most politically charged aspect of the negotiating round in fact involved regime review. The Convention required COP 1 to 'review the adequacy' of the commitments of the Annex I parties, and the negotiating parties as a whole were therefore required to negotiate a response to this.

Regime strengthening. The Kyoto Protocol negotiations: 1995–1997

While the adoption of regime-development decisions was a key feature of COP 1, the Conference will be best remembered for its decision to launch a new round of negotiations aimed at strengthening the commitments of industrialized countries, as a result of the above-mentioned review of adequacy. This decision, known as the Berlin Mandate after the host city of COP 1, was extremely hotly debated, but eventually established that the new round of negotiations would focus on setting 'quantified emission limitation and reduction objectives' (in other words, emission targets) *for industrialized countries*, but would not impose any new commitments on *developing countries*. The repercussions of this landmark decision, which steered the climate change negotiation process inexorably down a particular path, still reverberate strongly today. The deadline for these negotiations was set as COP 3.

While a degree of more routine regime development continued between COP 1 in 1995 and COP 3 in 1997, it was regime strengthening that overwhelmingly dominated the negotiation process. The Kyoto Protocol negotiations were conducted within the specially convened Ad Hoc Group on the Berlin Mandate (AGBM), a body open to participation by all parties, which met eight times over the 32-month period.

The geopolitical context for the Kyoto Protocol negotiations continued to be positive (Yamin and Depledge, 2004), with an actively engaged US administration, for example, although cracks were beginning to show. The realization on the part of many Annex I parties that curbing their emissions would be harder than they had hoped, and signs of a downturn in the global economy, cast a shadow over the negotiation round. The 'finale' of the negotiating round took place at COP 3 itself, where the Kyoto Protocol was eventually adopted more than 12 hours after the scheduled end of the conference. The Kyoto Protocol represents a considerable strengthening of the climate change regime, introducing quantitative emission targets for all industrialized countries, along with a series of flexibility mechanisms to help countries meet those targets and a set of more stringent monitoring, review and compliance procedures. Key features of the Kyoto Protocol are outlined in Box 3.2.[10] The Kyoto Protocol will use the same basic institutional structure as the Convention, with the UNFCCC COP serving as the meeting of the parties to the Kyoto Protocol, creating a body known as the 'COP/MOP'.

Source: Adapted from Yamin and Depledge (2004)

Box 3.2 Key elements of the Kyoto Protocol

- all parties: general commitments.
- Annex I parties: individual emission targets, adding up to a total cut of 5 per cent. Targets range from –8 per cent (most countries) to +10 per cent, and are listed in Annex B.
- emission targets:
  - cover carbon dioxide, methane, nitrous oxide, hydrofluorocarbons, perfluorocarbons and sulphur hexafloride, counted together as a basket.
  - also cover certain carbon sequestration activities – the land use, land-use change and forestry (LULUCF) sector – based on specific rules.
  - in most cases, use 1990 as a baseline.
  - must be met by the 'commitment period' 2008–2012.
- flexibility mechanisms[1]
  - joint implementation (JI), clean development mechanism (CDM) and emissions trading – can be used to help meet targets. Groups of countries can also meet targets jointly (so far, only invoked by the EU).
- stricter reporting and review procedures for Annex I parties.
- compliance system to address cases of non-compliance with the Protocol.
- regular reviews of commitments.

Note: 1 The term 'flexibility mechanisms' is not used in the climate change negotiations themselves, which refer instead to the more neutral 'Kyoto mechanisms'

Regime development. The post-Kyoto negotiations: 1998–2001

The conclusion of the Kyoto Protocol negotiations led directly to another negotiating round, with the decision adopting the Kyoto Protocol at COP 3 (decision 1/CP.3) calling for further work on the three flexibility mechanisms,[11] as well as the Protocol's provisions on LULUCF and on the impact of single projects.[12] In practice, however, the post-Kyoto negotiations did not really start until COP 4 in 1998, when the so-called Buenos Aires Plan of Action (BAPA) was adopted. In essence, the BAPA extended the remit of the post-Kyoto negotiations to encompass decisions on *all* the implementation rules of the Kyoto Protocol, akin to the Convention's prompt start process (described above). In parallel, the BAPA also launched negotiations aimed at further developing the Convention, including on such topics as financial assistance, technology transfer and the special circumstances of particularly vulnerable developing countries. Although the deadline was not so clearly articulated, the expectation grew that this mammoth negotiating round on both the Convention and the Protocol would result in a package deal on all these regime development issues at COP 6 in 2000.

Whereas the Kyoto negotiations were conducted in a single, specially convened forum (the AGBM), the post-Kyoto negotiations took place in the existing subsidiary bodies (see Figure 3.1), plus a 'joint working group' set up by COP 4 to conduct negotiations on a compliance system.

While it is indeed appropriate to characterize these negotiations under the BAPA as a 'regime development' round, it is also true that, at least in the case of their Kyoto Protocol component, they did involve an element of strengthening, or indeed weakening, as the specific rules concerning such provisions as the flexibility mechanisms and LULUCF would make a real difference to the effort required by Annex I parties to meet their emission targets. A degree of regime review also continued as the COP considered the national communications of Annex I parties, but very little importance was attached to this compared with regime development.

This negotiating round had, in effect, three finales. The first was at the scheduled end point of the negotiations, at COP 6 in The Hague in November 2000. This finale famously ended in ignominious collapse, with no package deal and no agreement on most of the individual issues under negotiation. COP 6, however, was resumed eight months later in Bonn in July 2001, where a political deal was reached on the so-called 'crunch issues' – the key controversial questions – in a decision known as the Bonn Agreements (decision 5/CP.6). In November that year, at COP 7 in Marrakesh, the 14 page Bonn Agreements were supplemented by the Marrakesh Accords (decisions 2-24/CP.7), incorporating over 210 pages of detailed rules and/or guidelines on all the BAPA issues. This marked the close of the post-Kyoto negotiations under the BAPA, although a few outstanding issues did carry over to COP 8 in 2002 and even COP 9 in 2003, notably specific technical aspects of the guidelines for reporting under the Protocol, and the more political question of the treatment of the forestry sector in CDM projects.

Two important events in the wider geopolitical environment helped shape the post-Kyoto negotiations (Yamin and Depledge, 2004). The first was the entrenchment of the widespread global economic slowdown, including dramatic crises in many of the 'Asian Tigers' and Latin American economies. The second was the 2000 US elections, which brought the Republican George W. Bush to the White House, rather than the environmentalist Democrat Al Gore. The elections took place just a week before the opening of COP 6 and, due to disputes over the counting of votes, it was unclear for much of the conference who the new US President would be. Before COP 6 resumed in July 2001, President Bush had repudiated the Kyoto Protocol (but not the Convention). The US rejection of the Kyoto Protocol, however, seemed to galvanize rather than dishearten most of the remaining Convention parties, who went on to adopt the Bonn Agreements, and later the Marrakesh Accords. COP 6 (part II), and to a lesser extent COP 7, concluded on a tide of optimism that the Kyoto Protocol could be brought into force and made to work, even without the participation of the US or its climate change ally, Australia.

Regime development. The post-Marrakesh negotiations (2002 to date)

The post-Kyoto negotiations have not been followed by a clear new phase. The climate change negotiation process has certainly continued, with a heavy agenda of work at COP 8 and COP 9. These negotiations have continued to develop the regime further, putting into practice the Marrakesh Accords, reviewing the implementation of commitments (through consideration of national communications), and tying up loose ends of the post-Kyoto negotiations. These regime development negotiations, however, have had no clear overall direction or goal, unlike previous negotiating rounds. There are two reasons for this. Firstly, the natural result of the adoption of the Marrakesh Accords, and their conclusion of the details of the Kyoto Protocol's rules, was for the regime to enter an implementation phase, with Annex I parties focused on meeting their emission targets and otherwise putting the Protocol into practice. This focus on implementation was blurred to a large extent by the delay in securing sufficient ratifications for the Kyoto Protocol to enter into force, notably prevarication by the Russian Federation. However, the entry into force of the Protocol in February 2005 promises to launch a true implementation phase. Another related reason why the climate change negotiations have seemed to enter into a period of relative stagnation is the inability of the parties to launch a new regime-strengthening negotiating round. While most Annex I parties have long called for new negotiations to strengthen the commitments of developing countries, the resistance of developing countries to doing so shows no sign of diminishing, and indeed has hardened with the delay in entry into force of the Kyoto Protocol and the extremely modest progress made by major emitters in stemming their rising GHG emissions.

The hope is that the entry into force of the Kyoto Protocol, and its effective implementation by Annex I parties, may help to break the political deadlock.

The challenges of the climate change negotiations

As discussed in Chapter 2, two defining characteristics of global intergovernmental negotiations are complexity and inequality. The climate change negotiations exemplify these characteristics absolutely, exacerbated by the malign nature of climate change itself, as outlined above. This raises challenges to the successful unfolding of the climate change negotiations.

Table 3.1 *The three main phases of the climate change negotiations*

Period	Phase and meeting held	Summary
1995–1997	Kyoto Protocol negotiations 1995 – COP 1; AGBM 1–2; SBSTA/SBI 1; AG13 1 1996 – COP 2; AGBM 3–5; SBSTA/SBI 2–4; AG13 2–3 1997 – COP 3; AGBM 6–8; SBSTA/SBI 5–7; AG13 4–5	Negotiations launched by the 'Berlin Mandate' ar COP 1 that led to the adoption of the Kyoto Protocol at COP 3. Negotiations on the Kyoto Protocol conducted in the AGBM. The SBSTA and SBI also met to develop the Convention provisions further. The ad hoc group on Article 13 (AG13) was convened to negotiate a multilateral consultative process (ultimately unsuccessfully).
1998–2001	Post-Kyoto negotiations 1998 – COP 4; SBSTA/SBI 8–9; AG13 6 1999 – COP 5; SBSTA/SBI 10–11 2000 – COP 6; SBSTA/SBI 12–13 (parts I and II). 2001 – COP 6 (part II), COP 7; SBSTA/SBI 14–15	Negotiations on the details of the Kyoto Protocol, as well as the implementation of the Convention, based on the 'Buenos Aires Plan of Action' adopted at COP 4. These negotiations, conducted in the SBSTA and SBI, failed to conclude as scheduled at COP 6. COP 6 (part II) reached agreement on a political deal covering the key political issues – the 'Bonn Agreements'. The more technical details were agreed at COP 7 in the 'Marrakesh Accords'.
2002–	Post-Marrakesh negotiations 2002 – COP 8; SBSTA/SBI 16–17 2003 – COP 9; SBSTA/SBI 18–19	Continuing negotiations on the routine development of the Convention and Kyoto Protocol under the subsidiary bodies.

In common with all global processes, the climate change negotiations are rendered complex by the many parties involved. The UNFCCC states that membership of the regime is open to any State member of the UN, or its specialized agencies or the International Atomic Energy Agency. The number of parties to the UNFCCC has risen from 118 at the time of COP 1, to 189 at COP 9. More than 80 per cent of these parties are represented at the negotiating sessions themselves. Since COP 3, every COP session has enjoyed participation by more than 150 parties, with a peak of 179 at COP 6 (part I). Each party is represented by a delegation. The size of delegations ranges from just one or two members from many smaller developing countries to more than 20, and sometimes many more for 'mega delegations' such as those regularly fielded by the US, Canada or Japan. The number of *individual* country delegates at negotiating sessions has exceeded 1300 at every COP session since COP 3, and has hovered between 600–800 at every subsidiary body session. The high point was at COP 3 with nearly 3000 individual country delegates.

The potential for a large range of preferences and proposals inherent in a global negotiation is certainly realized in the climate change context by the diversity of specific national circumstances relative to the problem, superimposed on broader inequalities of wealth, power and stage of development. Differing degrees of vulnerability to climate change impacts, economic dependence on the sale of fossil fuels, and emissions per capita or per unit of GDP, are major variables that split both the developed and the developing countries (see Box 3.3). As noted in Chapter 2, however, intangible values and principles are often as important as tangible perceived interests in shaping a negotiator's position. In the context of climate change, these values and principles tend to revolve around the following (see also Box 3.3):

- the importance attached to environmental protection
- attitudes to multilateral cooperation in general, including levels of trust and geopolitical relations with other nations
- broad perceptions of the merits of different policy instruments (notably attitudes to market mechanisms, for example, emissions trading versus government regulation).

These themes are not exclusive to climate change, but also tend to feature prominently in debates on other global environmental issues.

The substantive inequalities in wealth and power among parties to the climate change regime are reflected in procedural inequalities in terms of negotiating capacity, which crudely correlate with delegation size. Small delegation size is usually both a cause and symptom of a broader paucity of resources, including a narrower breadth of expertise and lack of preparation time for negotiating sessions. One developing country interviewee recalled: 'Our delegation in Kyoto, there were two of us. And the first person... didn't really know about the negotiations. So I was alone, running backwards and forwards, making contacts... It was a lot of work ...It's difficult.' The difficulties faced by smaller delegations in participating effectively in the negotiations are illustrated powerfully by Figure 3.2 below.

Source: ECO (1997b)

Figure 3.2 *International treaty-making: A model for global democracy*

Akin to almost all global negotiation processes, a number of negotiating coalitions are active in the climate change negotiations, and these play an important role in shaping the dynamics of the climate change process (Yamin and Depledge, 2004). Some of these coalitions are specific to the climate change context, while others are at work also in other arenas (see Box 3.3). Some are founded more on values and principles, often with a common historical link, while others are based more on shared national circumstances. The coalitions also have different degrees of cohesion, with some speaking with one voice, and others sticking to more informal information sharing. The pattern of coalitions in the climate change regime has not remained static, but has evolved over time, with new coalitions forming and others disbanding in response to developments in the regime. This is a case where there is a clear iterative relationship with the organization of the negotiation process. As we shall see in Chapter 9, recent trends in the way in which negotiating arenas in the climate change process have been organized have triggered an increase in the number of active negotiating coalitions.

The complexity of many diverse parties and their varying preferences are mirrored, and accentuated, in the NGOs that are active in the climate change process. As discussed further in Chapter 14, these cover a very wide range of interests, including environmental NGOs, business and industry representatives, indigenous peoples, faith groups and academics. While NGOs do not actually negotiate, they contribute to the complexity of the process in many ways. NGOs add to the (contradictory) pressures on parties that influence negotiating positions, they require channels through which to communicate their views and feed into the negotiation process, and the sheer larger number of people (who must also be treated differently to representatives of parties) creates logistical challenges (e.g. demand for meeting rooms, seating, security).

A different group of non-state organizations are the intergovernmental

Box 3.3 Major coalitions in the climate change regime and their basic positions

Group of 77 and China: The G-77 is the main coalition of developing countries in the UN system. China often allies itself with the Group, which numbers 132 members. The G-77 was formed in 1964 during negotiations on a 'New International Economic Order' explicitly to counter the might of the developed world. This history, combined with persistent inequalities and the development *and* equity dimensions of climate change, are carried through to a deep North–South divide that permeates every aspect of the climate change negotiations. The G-77 has always perceived climate change as a development issue, invoking equity as the fundamental principle for addressing it. A basic position of the G-77 is that the developed countries must cut their own emissions, before requiring the developing countries to do so. The G-77 also emphasises the need for financial assistance and technology transfer, informed by a belief in multilateral cooperation. Many G-77 members find themselves disadvantaged in bilateral exchanges, and value the protection that rule-based institutions can offer. They have tended to be suspicious of market mechanisms, seeing these as possible loopholes for evading commitments. North–South relations in the climate change regime are indeed characterized by generalized mutual mistrust.

While G-77 members broadly share common principles, their national circumstances vary considerably. This is reflected in the other groups that operate within the G-77.:

- *Alliance of Small Island States*: AOSIS was formed in 1990 specifically to lobby on climate change. It now comprises some 43 low-lying or small island states, most of them also G-77 members. AOSIS members are among the most vulnerable to climate change, and have argued forcefully for strong commitments to reduce emissions.
- *Least Developed Countries*: The LDCs, who operate throughout the UN, are highly vulnerable to desertification, drought and extreme weather brought on by climate change, exacerbated by poverty and lack of resources. LDCs are located in Africa (mostly) and Asia, with some also members of AOSIS.
- *Organization of Petroleum Exporting Countries*: Fearing the economic repercussions of a decline in demand for oil, OPEC has opposed strong mitigation action. Some members have even been accused of obstructing the negotiations. It does not negotiate as a group, but coordinates very effectively informally.

European Union: The 25 member EU almost always speaks with one voice, despite differences in the national circumstances of its members. The EU views itself as an environmental leader, supporting strong commitments and with a traditionally lukewarm attitude towards market mechanisms. The EU is also generally more comfortable with government regulation and multilateral cooperation than most of the Umbrella Group. The EU has not always translated its leadership aspirations into concrete influence, due to internal divisions and cumbersome internal procedures (see Gupta and Grubb, 2000). The EU did, however, exercise important leadership in upholding the Kyoto Protocol in the wake of the US repudiation.

Economies in Transition: The EITs, 10 of which are now EU members, experienced deep emission cuts in the early 1990s, due to economic collapse. Prior to this, the EITs were among the most carbon intensive economies in the world (see Househam et al, 1998). The new EU entrants, who previously negotiated as the Central Group-11, traditionally supported the EU. The Russian Federation and Ukraine, however, are in the Umbrella Group.

Umbrella Group: This Group's members share similar values and principles in the climate change negotiations, centred on the pursuit of flexibility and cost-effectiveness. Their national circumstances, however, are very different. Japan, New Zealand, Norway and Iceland, for example, have far lower emissions per capita and per unit of GDP than the geographically larger and more carbon intensive economies of Australia, Canada and the US, or the Russian Federation and the Ukraine. They are also politically diverse. The US and Australia have repudiated the Kyoto Protocol, but the others have ratified. Some of its members (e.g. Canada, Norway) are strong supporters of multilateral cooperation on the environment, in contrast with the US under the George W. Bush Administration at the time of writing. This explains why the Umbrella Group is only a loose coalition, which rarely negotiates as a single entity.

organizations (IGOs) such as the International Energy Agency (IEA), as well as UN bodies, such as the UN Environment Programme (UNEP) and the UN Development Programme (UNDP). Representatives from these IGOs focus mostly on providing specialist advice to their members, showcasing their climate change related activities, or monitoring developments of interest to them (Yamin and Depledge, 2004). The climate change regime also tends to attract a high media presence, which fluctuates depending on the stage of negotiations.

An important trend in the post-Kyoto era contributing to the complexity of the climate change negotiations has been an increase in the number of different issues on the negotiating agenda. This can be attributed to several factors. Firstly, the awareness of countries of their particular national circumstances and interests relative to climate change has risen over time, so that more specific issues are being placed on the table. Secondly, the adoption of the Kyoto Protocol and the Marrakesh Accords created a whole new gamut of mechanisms and procedures that must be overseen and monitored by the negotiating parties; this, again, was a product of the differing positions of the negotiating parties, which all had to be accommodated. Thirdly, there is inherent difficulty in ever closing an agenda item; once one has been launched, this creates a new community of people with an interest in keeping it alive. Fourthly, the rules for the conduct of business are expansive when it comes to admitting and keeping items on the (at least provisional) agenda. This greater number of issues has encouraged specialization, so that there are fewer delegates with the big picture of the whole negotiation process.

A key challenge for the climate change negotiations since their inception has been a strong tendency to competitiveness, rather than cooperation, among the negotiating parties. This is partly the result of the high political stakes of climate change, including concerns over national economic interests and competitiveness, as well as the long time horizon of the problem, which has led to a focus on short-term costs rather than on benefits that would only accrue in the future. The tendency to competitiveness is also a product of the North–South divide to the negotiations, where the imperative of global cooperation struggles against a history of mistrust and differing perceptions of the problem.

The most extreme case of competitiveness is that of the OPEC parties, whose negotiating positions appear to be based on a stronger readiness to obstruct than to cooperate given their fears over the economic impact of a carbon-constrained world. The conflictual nature of relationships between the negotiating parties is in turn reflected in the deep chasm between environmental NGOs on the one hand and the fossil fuel industry NGOs on the other. Putting these elements together, a particularly unenviable feature of the climate change negotiations is the presence of an alliance of obstructionist forces, including carbon intensive lobby groups and a small number of eloquent negotiating parties, whose goal for the negotiation process is simply to slow it down. At a fundamental level, therefore, the negotiations have to overcome the fact that an influential minority of negotiating parties, assisted by well funded non-state organizations, in fact do not support the negotiation process and seek to block an outcome.

The organization of the climate change negotiations

In the following chapters, we will examine how the climate change negotiation process has been organized to try to face up to, and overcome, its challenges. As discussed in Chapter 2, the organization of a global intergovernmental negotiation is typically founded upon a hierarchy of formal and informal procedural rules and practices. These rules and practices are then used by the organizers of the negotiation process – the presiding officers, bureau and secretariat – as the basis for their organizational decisions. The origins and nature of these rules and practices in the climate change context are outlined below.

Founding texts

The Convention itself establishes the institutions of the climate change regime (see Figure 3.1) and sets out basic procedural rules relating to the conduct of business and decision-making. It establishes, for example, procedures for considering and adopting proposed amendments, new annexes and protocols to the Convention, and also for holding sessions of the COP, including the admission of observers. In its Article 18, the Convention sets out the fundamental rules for participation in the climate change negotiations, that is, on a one-party-one-vote basis. The procedural rules included in the Convention are similar to those in other multilateral environmental agreements (MEAs),[13] and are also carried over, mutatis mutandis, to the Kyoto Protocol.

Rules of procedure

The Convention called on the first COP to 'adopt its own rules of procedure as well as those of the subsidiary bodies' (Article 7.3). Due to disagreement over the voting rule to be applied for taking substantive decisions, the draft rules of procedure have never been adopted (Yamin and Depledge, 2004). However, given that the bulk of the draft rules of procedure are not in contention, the draft rules (see FCCC/CP/1996/2) have been 'applied' at each session of the COP and subsidiary bodies without controversy, except for the rule on voting (rule 42). Each COP President has been charged with consulting to resolve the impasse. However, since the adoption of the Kyoto Protocol, the COP President's consultations have been cursory at best, or have not taken place at all, and the application of the draft rules of procedure at each session has become unchallenged established practice. For this reason, and in the interests of readability, this book will refer simply to the 'rules of procedure'.

The rules of procedure, which include a broad set of rules for the functioning of the regime, are largely unexceptional in the multilateral environmental arena. The Kyoto Protocol, for its part, specifies that it will apply the Convention's rules of procedure, 'except as may be otherwise decided by consensus' by the COP/MOP (Kyoto Protocol Article 13.5). At the time of writing, the understanding was that COP/MOP 1 would indeed apply the COP's rules of procedure.

Supplementary decisions

The COP has adopted a small number of formal decisions relating to the organization of the negotiation process. Decision 18/CP.4, for example, sets out procedures for the participation of observers in contact groups, while other decisions have been taken to better define the division of labour between the two subsidiary bodies, and to try to control the increase in volume of documentation. More commonly, however, the SBI will adopt organizational conclusions on the forthcoming COP, including, for example, on how to involve ministers and on the basic structure of the Conference.

Informal practices

The organization of the climate change negotiation process is replete with informal practices that have developed over time. The organizers of the negotiation process have indeed been highly active in seeking to interpret the regime's procedural rules, and improvise based on those rules, in order to respond to the needs of the negotiation process and organize it better. The informal practices of the climate change negotiations are sometimes difficult to pin down, precisely because they are unwritten. They include, for example, the 'no more than two meetings' rule, whereby only two official negotiating groups may meet at any one time, along with the practice of appointing two Co-Chairs to preside over informal groups, one each from an Annex I and non-Annex I party.

Wider institutional setting

The institutional setting of the UNGA plays a very influential role in the organization of the climate change negotiation process. This includes, for example, formal rules surrounding the provision of interpretation and translation, along with informal practices. The 'no more than two meetings' rule discussed above, for example, is carried over from the wider UN system.

Summary and concluding remarks

The climate change negotiations face a number of difficult challenges, arising from the malign nature of climate change, combined with the complexity and inequality that characterize global intergovernmental negotiations and are particularly prevalent in the climate change context. We now turn to analyse how organizational elements have been used in the climate change negotiations to try to meet these challenges, beginning with an exploration of the role of the organizers of the negotiation process – the presiding officers (Chapter 4), bureau (Chapter 5) and the secretariat (Chapter 6).

Presiding Officers

... the real task of being the Chair... includ[es] almost every kind
of role, between on the one extreme being the spiritual adviser or
psychologist (weeping delegates!), through being a manipulator
or seducer, to the other extreme of being a dictator (Veit Koester
(Chair, negotiations on the Cartagena Protocol on Biosafety), in
Bail et al (2002, p46)).

Introduction

Presiding officers play a central role in any global negotiation. Together with the
secretariat, presiding officers have overall responsibility for the smooth conduct of
the negotiations, and as such their actions shape every facet of the organization of
the negotiation process. A hierarchy of presiding officers exists in the climate
change regime, corresponding to the three layers of negotiating arenas over which
they preside (see Chapter 9, also Figure 3.1). At the top of the apex sits the
President of the COP, typically a Minister. Then, at the second institutional layer
of the regime, come the Chairs of the SBSTA and the SBI. These three officers –
COP President and subsidiary body Chairs – are formally elected by the parties to
the regime, with their basic terms of reference (e.g. length of service) governed by
the rules of procedure. Occupying the third institutional layer of the climate
change regime are the Chairs of informal groups established by the COP or
subsidiary bodies to negotiate specific issues for one session only.

In this chapter, we explore the roles of all these presiding officers, the ways
in which they can contribute to the effective organization of the negotiating
process, and the qualities and skills required for them to do so.

Mandate, functions and appointment

The formal mandate and functions of the COP President, which also apply
to the subsidiary bodies,[1] are set out in the rules of procedure of the climate
change regime. Although informal groups are not expressly provided for
under the rules, the general description of a presiding officer's role also
pertains to their Chairs.

Mandate and functions

The foundation stone for the work of a presiding officer is *impartiality*. According to the rules, the President 'shall not...exercise the rights of a representative of a Party' (Rule 22.3). Building on this foundation, the presiding officer's mandate rests on a delicate balance between *authority and deference* relative to the parties. On the one hand, the rules state that the presiding officer 'shall have *complete control* over proceedings and over the maintenance of order thereat', on the other hand, they caution that s/he '*remains under the authority* of the Conference of the Parties' (Rule 23.1, 23.3). Presiding officers are therefore granted a fragile mandate, which must be managed with care.

The rules outline several specific functions for presiding officers, which are common in the multilateral arena. These include the following (Yamin and Depledge, 2004):

- declare the opening and closing of sessions, preside over meetings and ensure the observance of procedural rules (rule 23.1)
- accord the right to speak (rules 6, 23, 32)
- put questions to the vote and announce decisions (rule 23.1)
- rule on points of order (with the possibility of challenge) (rules 23.1, 34)
- draw up the provisional agenda for each session, with the secretariat (rule 9).

These limited formal functions, however, belie a potentially much richer informal role that is accepted – and often expected – of a presiding officer in promoting the reaching of agreement, in other words, the exercise of process-oriented leadership. In a letter to the COP 2 President, former Executive Secretary Zammit Cutajar (the head of the secretariat) explained, 'the political functions of the President are not set out in writing, but are expected to include leadership in seeking consensus' (Zammit Cutajar, 1996). In a study of numerous environmental regimes, Underdal similarly concludes that, although the roles of 'conference presidents and committee chairs are most often quite narrowly circumscribed in terms of formal authority...nonetheless incumbents often succeed in using such roles as an important basis for exercising influence' (Underdal, 2002, p27 and note 28). Such an informal role opens up a wide, though not neatly defined, potential space for action. According to Lang, 'the informal powers of a presiding officer are neither codified nor really limited – whatever serves the purpose of the conference and is accepted by the participants may be undertaken' (Lang, 1989a, p39). However, he goes on to warn that 'the extent to which a presiding officer may use most of his informal powers depends on the permissive consensus of the participants...the powers of a presiding officer are extensive and fragile at the same time'.

Although the general nature of their roles is the same, the actual tasks of different types of presiding officers will vary in practice, according to the functions of the body over which they preside. The COP President tends to have a much more ceremonial role, as almost all decisions will already have been reached in the subsidiary bodies or informal groups before they are presented to the COP plenary (see Chapter 9). Although COP Presidents can choose to play an active role in promoting agreement, in practice only a small number have done so in any truly consequential way, and in some cases their efforts have even backfired. The subsidiary body Chairs, for their part, as the presiding officers of the main working bodies of the regime, have a very important role to play in pushing for an overall agreement on the issues under their purview. This is also the case for informal group Chairs, although these will usually be dealing with single issues. Informal group Chairs also know that they can, in the last resort, appeal back to the COP or subsidiary bodies if agreement cannot be reached.

Of all the organizational elements considered in this book, the role of the presiding officer is viewed in the literature as holding the greatest potential to impact on a negotiation (e.g. see Kaufmann, 1989; Young, 1991; Schermers and Blokker, 1995). This view is backed up by numerous empirical cases, where the actions of a skilful presiding officer contributed to a successful negotiation, while less judicious chairing placed obstacles to agreement. Lang suggests that, while the contribution of even a particularly activist presiding officer is unlikely to exceed 'ten percent of the total impact of all negotiators ... in some instances ... this ten percent makes the difference between success or failure' (Lang, 1989a, p39).

Appointment

COP President
The rules specify that the position of COP President is 'normally subject to rotation' among the five UN regional groups - Africa; Asia; Central and Eastern Europe; Latin America and the Caribbean; and Western Europe and Others (Rule 22.1). An informal tradition has also developed, based on common practice in the UN system, whereby the venue for the COP usually rotates among the five regional groups together with the position of President. Any country may offer to host a COP. Although the offer must be formally examined by the SBI and approved by the COP, it is unlikely that an offer will be refused. If no one offers to host, the session is held at secretariat headquarters in Bonn (Germany), and regional rotation continues to apply to the presidency. The practice in the climate change regime is for the position of COP President to be held at ministerial level, usually by the minister for environment.

The President is formally elected at the opening of the COP session that s/he will preside over. In practice, the process of election is a formality as, given that the presidency goes hand in hand with the COP venue, or at least with regional rotation, the identity of the President is typically known long before the COP session opens. The President remains in place until the next session of the COP, presiding over any inter-sessional meetings (see Chapter 9), then hands over to the following COP President at its opening. The Presidents that have served in the climate change regime to date are listed in Table 4.1 below.

Subsidiary body Chairs
The subsidiary body Chairs are elected as part of the 11-member COP Bureau (see Chapter 5). Although the composition of the Bureau as a whole is subject

Table 4.1 COP *Presidents to date*

COP	Dates	Venue	President	Country	UN Region
COP 1	28 March–7 April 1995	Berlin	Merkel	Germany	Western Europe and Others Group
COP 2	8–19 July 1996	Geneva	Chimutengwende	Zimbabwe	Africa
COP 3	1–11 December 1997	Kyoto	Ohki	Japan	Asia
COP 4	2–14 November 1998	Buenos Aires	Alsogaray	Argentina	Latin America and the Caribbean
COP 5	25 October–5 November 1999	Bonn	Szyszko	Poland	Central and Eastern Europe
COP 6	13–24 November 2000	The Hague	Pronk	Netherlands	Western Europe and Others Group
COP 6 Part II	13–27 July 2001	Bonn	Pronk	Netherlands	Western Europe and Others Group
COP 7	29 October–9 November 2001	Marrakesh	El Yazghi	Morocco	Africa
COP 8	23 October–1 November 002	New Delhi	Baalu	India	Asia
COP 9	1–12 December 2003	Milan	Persányi	Hungary	Central and Eastern Europe
COP 10	6–17 December 2004	Buenos Aires	[to be elected]	Argentina	Latin America and the Caribbean

Source: Adapted from Yamin and Depledge (2004)

to rules regarding regional distribution, the position of subsidiary body Chair is not. An informal understanding prevails, however, that the position of Chair will rotate fairly among the regional groups over time. As with the COP Bureau as a whole, the positions of subsidiary body Chairs are usually decided upon through the (sometimes highly contentious) process of regional nomination behind the scenes, and election is normally a formality. The subsidiary body Chairs, along with all Bureau members, are permitted to serve two terms, and usually do (Yamin and Depledge, 2004).

According to the rules, the Chairs of *new* subsidiary bodies are elected by the subsidiary bodies themselves (Rule 27.5). This applied, for example, to the ad hoc group on Article 13 (AG13) that met 1995–1998 to negotiate (ultimately unsuccessfully) a multilateral consultative process under the Convention, along with the Joint Working Group (JWG) that met 1999–2000 to negotiate the Kyoto Protocol provisions on compliance. However, in the case of the AGBM, the ad hoc body convened to negotiate the Kyoto Protocol, its Chair, Raúl Estrada-Oyuela, was designated by the COP on the Group's establishment, enabling preparations for the AGBM's first session to start immediately in a context of procedural certainty. The subsidiary body Chairs that have served to date are listed in Table 4.2 below.

Informal group Chairs

The Chairs of informal groups are invited to serve by the presiding officer of their convening body, usually the SBSTA or SBI, or less frequently the COP. The Chairs of informal groups are appointed for one session only, but may be re-appointed at subsequent sessions if the informal group is reconvened. During the post-Kyoto negotiations, for example, several informal groups – known as contact groups (see Chapter 9) – convened to work on the major issues under negotiation,[2] acquired a momentum of their own as their mandates were renewed and their Chairs re-appointed session after session. The performance of these contact group Chairs in steering the negotiations therefore became as important, if not more important, than that of the subsidiary body Chairs and COP Presidents.

The subsidiary body Chairs will usually give some thought prior to the session on who could chair the informal groups they are thinking of convening, and will discuss possible names with the secretariat. In drawing up a list of possible informal group chairs, and in keeping with the foundation of impartiality that must underpin the work of any presiding officer, the subsidiary body Chairs will seek to maintain regional balance among the full slate of informal group Chairs, and to ensure that the chairing will be perceived as impartial. In this respect, a practice has emerged post-Kyoto of appointing Co-Chairs for contact groups,[3] one each from an Annex I and a non-Annex I country, to try to make sure that both sets of parties feel their interests will be given due attention. This practice, however, has the disadvantage of entrenching the divide between Annex I and non-Annex I parties, while doubling the demand for effective Chairs. Several secretariat interviewees pointed to a 'scraping of the barrel', whereby inexperienced individuals are invited to chair, sometimes ineffectively. In many cases, the practical result is

Table 4.2 *Chairs of the SBSTA and SBI to date*

Term of office	SBSTA	SBI
Elected at COP 1 Served SBSTA/SBI 1–7	Tibor Faragó (Hungary)	Mahmoud Ould El Ghaouth (Mauritania)
Elected at COP 3 Served SBSTA/SBI 8–10	Chow Kok Kee (Malaysia)	Bakary Kanté (Senegal)
Elected at COP 5 Served SBSTA/SBI 11–15	Harald Dovland (Norway)	John Ashe (Antigua and Barbuda)
Elected at COP 7 Served SBSTA/SBI 16–17	Halldor Thorgeirsson (Iceland)	Raúl Estrada-Oyuela[1] (Argentina)
Elected at COP 8 Served SBSTA/SBI 18–19		Danieka Stoycheva (Bulgaria)
Elected at COP 9 Served SBSTA/SBI 20–21	Abdullatif S. Benrageb (Libya)	
	Ad hoc bodies	
AGBM (1995–1997)	Raúl Estrada-Oyuela (Argentina)	
AG13 (1995–1998)	Patrick Széll (UK)	
Joint Working Group on Compliance (1999–2000)	Harald Dovland (Norway)	Espen Ronneberg (Marshall Islands) (to October 1999) Tuiloma Neroni Slade (Samoa)

Source: Adapted from Yamin and Depledge (2004)
Note: 1 Estrada only served one term as SBI Chair, as he had previously served an additional term on the COP Bureau as a Vice-President, and therefore was not eligible to stand again

that one active Chair emerges, while the other Chair remains largely a figurehead.

In order to avoid controversy over the appointment of informal group Chairs, at several sessions (e.g. at COP 6 (part II)), they have been chosen from among COP Bureau members. This is not always possible, however, as there is no guarantee that COP Bureau members will hold the necessary skills. Another means of avoiding controversy has been to invite the groups of Annex I and non-Annex I countries to nominate their own Co-Chairs, a practice that again emerged during the post-Kyoto negotiations. While this has the advantage of ensuring legitimacy, it also runs the danger that a Co-Chair will be nominated not because of his/her aptitude, but because of his/her political acceptability or because of strong pressure from a powerful party. Although it is, of course, difficult to deduce the motives of parties, one possible example of this occurred at the 13th session of the subsidiary bodies (SBSTA/SBI 13 in September 2000), when a representative of Saudi Arabia was nominated by the G-77 as the non-Annex I Co-Chair of the contact group on adverse effects,[4] despite the extreme and uncompromising position of that country on the topic.

The COP 6 President pioneered a new practice for the finale of the post-Kyoto negotiations of asking *ministers* to chair the final informal groups. At COP 7, the President similarly adopted a novel approach of asking two ministerial colleagues, from Switzerland and South Africa, to serve as 'co-facilitators' for the final stages of negotiations. The term 'facilitator', which appears to have been introduced by COP 6 President Pronk, was used in preference to 'chair' at both COP 6 (part II) and COP 7 because of its more informal connotations. As one secretariat interviewee put it, the term 'is meant to invoke feelings of soft, consensual confidence building... "Chair"... is harsher'. Using ministers in this way, however, while appropriate when high-level political decisions must be taken, does require understanding of their differing strengths and dynamics compared with officials, as discussed further in Chapter 13.

The roles of presiding officers

There are several different ways in which presiding officers can exercise process-oriented leadership to help overcome the challenges faced by global negotiations.

Strategic organization of the negotiation process

Central to the exercise of leadership by the COP President and subsidiary body Chairs, as presiding officers of the Convention bodies, is overall strategic organization of the negotiations, usually in partnership with the secretariat. Examples of such strategic organization relating to the conduct of business and decision-making, use of different negotiating forums, choice of negotiating texts, management of time, and involvement of NGOs and ministers, appear throughout the individual chapters that follow.

In terms of complex negotiating rounds that span several sessions, such as the Kyoto and post-Kyoto negotiations, the ability of presiding officers to contribute to the strategic organization of the round depends to a large extent on the appointment of a single clear leader throughout that round. In this respect, it is instructive to compare the Kyoto and post-Kyoto negotiations. The Kyoto negotiations were led by a single presiding officer, Chair of the AGBM Raúl Estrada-Oyuela, who stayed in office throughout the negotiating round, and was designated to chair the Committee of the Whole which conducted the final negotiations on the Protocol at COP 3 in Kyoto. This continuity was critical; Estrada was able to plan the negotiations strategically, develop an in-depth understanding of the negotiation process and build relationships with delegates, while parties became accustomed to his chairing style, thus stabilizing expectations and enhancing the efficiency of the process. The Chairs of the SBSTA and SBI, and Presidents of COP 2 and COP 3, played a supporting role, but it was clear that Estrada was, in practice, in charge.

The post-Kyoto negotiations were much more fragmented. Despite a unified framework provided by the Buenos Aires Plan of Action (BAPA), responsibility for the issues under negotiation was divided between three bodies (the SBSTA, SBI, and JWG – see Chapter 9) involving, throughout the negotiation process, a total of seven presiding officers for these bodies. The final negotiations at COP 6 (part I) were handed over to the eighth presiding officer, the Dutch COP President Jan Pronk. President Pronk did his best to bring the process together and exercise overall leadership, including convening informal consultations prior to COP 6 and preparing a single paper dealing with all the issues as a whole (see Chapters 9 and 11), but it was arguably too late, as he had no time to develop strong working relationships with, or the confidence of, parties. Ott (2001, p285) encapsulated the situation thus: 'Estrada...had presided over the negotiations on the Kyoto Protocol for several years, whereas the Dutch minister was a newcomer'. The absence of a single leader throughout the post-Kyoto negotiating round thus hampered the rise of strong process-oriented leadership so that the negotiations lacked an overall sense of direction.

Conduct of meetings

Successful chairing of meetings lies at the heart of the work of any presiding officer. In the case of a global negotiation, where meetings involve many parties from different linguistic, cultural and political backgrounds, the challenges are enormous. A key challenge is to reconcile the demands, and indeed rights, of all parties to be heard (procedural equity, transparency) with the imperative of ensuring that each meeting concludes its work in the time allotted (efficiency). Achieving such a balance requires knowing how to curb both long-winded but well-meaning speakers, and deliberate obstructionists and filibusters. Linked to this, a key goal must be to nurture and sustain a cooperative atmosphere among all the parties, however difficult the negotiations get, and try to overcome the tendency to competitive behaviour.

Chairing skills are especially important for the presiding officers of the subsidiary bodies and informal groups, which are the bodies that carry out the bulk of the work in the climate change regime. Indeed, most interviewees commented that COP Presidents have very rarely possessed the necessary experience or ability to manage spontaneous, genuine bargaining or even discussion in a plenary meeting. COP meetings are therefore always tightly choreographed in advance by the secretariat and COP Presidency to avoid surprise interventions and minimize unplanned debate (Yamin and Depledge, 2004).

There was widespread agreement among interviewees over a small number of presiding officers with strong chairing skills. The most high profile among these was Raúl Estrada, who supplied the Kyoto Protocol negotiation process with highly skilled chairing, which helped to maximize the efficiency of the negotiations, while promoting a cooperative atmosphere and sustaining procedural equity. According to former Executive Secretary Zammit Cutajar, Estrada possessed 'the sheer ability to chair a big meeting I have never seen before or since' (interview).

Estrada chaired the negotiations in a largely informal, personal and interactive manner, which helped to build up a sense of personal involvement and ownership of the negotiation process among parties. For example, he referred to the parties in the AGBM as 'we' and made a point of calling on speakers by their names, not just their countries. When he was not personally acquainted with a delegate seeking the floor, he would sometimes ask the secretariat to find out his/her name. Estrada was certainly extremely adept at reading the sense of the room and adapting his chairing style accordingly. As one interviewee noted, 'he had this incredibly disarming ability...to be able in a rather broad brush way to read the personalities he was dealing with, to know...whether to encourage their ideas, whether to allow them to ramble on or to rein them in'. An important tool for Estrada was the use of humour. Jokes, often at his own expense, enabled him to defuse tense situations, generate a more cooperative atmosphere, push through sensitive decisions, and disarm potentially difficult delegates who he feared were about to raise objections.

Humour and use of allegory have also been useful tools for other Chairs. SBI Chair Bakary Kanté kept up a constant flow of humorous remarks, which deflected attention from the difficulties he faced, as a non-English speaker, in chairing meetings. Chow Kok Kee, Chair of the contact group on flexibility mechanisms throughout the post-Kyoto negotiations (and previously SBSTA Chair), was renowned for likening the consideration of very complex negotiating texts to a 'walk through a rose garden', on which he would lead the delegates. This fostered a sense of community among contact group participants – albeit sometimes provoked by slight irritation at the repeated use of the metaphor – and helped launch often contentious negotiations in a more good-natured manner. Differing cultural backgrounds, however, means there are potential pitfalls to watch out for when using humour in global negotiations. The same Chair Chow, for example, when chairing the SBSTA, caused some disquiet by repeated references to 'girl power' when he appointed two female Co-Chairs to preside over a contact group.

Another case of highly effective chairing of large meetings cited by many interviewees was that of SBSTA Chair Halldor Thorgeirsson. Thorgeirsson achieved a particularly good balance between procedural equity and efficiency, managing to engage all delegates in an open manner, but at the same time keeping meetings on track and, as one interviewee put it, not 'letting people get carried away with their interventions'. Another interviewee commented on his political acumen, whereby he always found the right words to deal with both 'stupid questions' and obstructionism, while maintaining high awareness of political sensitivities.

Consensus building

As impartial, authoritative individuals, it is expected that presiding officers will play an important role in consensus building among the negotiating parties. In addition to making substantive proposals (see below), this will usually involve convening small, behind the scenes meetings of key players to discuss their differences and 'knock heads together'. Presiding officers are often able to extract concessions that parties are unwilling to make directly to each other, as parties can 'save face' by attributing their concession to the intervention of the presiding officer, and their desire to help him/her in reaching agreement, rather than any climb down on their part (see also Chapter 11). Here, there is often interplay between the presiding officers at different levels in the negotiation process. Contact group Co-Chairs, for example, may appeal to the subsidiary body Chair to intervene in the contact group negotiations on a particularly difficult issue, perhaps by inviting the main players to a private meeting.

There are innumerable examples of such consensus building by presiding officers. The intervention of SBSTA Chair Thorgeirsson, for example, was pivotal throughout the negotiations on the review of IPCC Third Assessment Report (TAR) when these became blocked in the contact group, while SBI Chair John Ashe specialized in 'knocking heads together' in private encounters, often with great success.

Presenting substantive proposals

Presiding officers may present substantive proposals on their own authority when they think this could help forge a consensus, and are indeed expected to do so. Depending on the nature of the negotiation, these can take the form of a comprehensive Chair's text (see Chapter 11), or language on a single issue or even a single phrase. The extent to which presiding officers will actually do this will vary considerably, depending on their knowledge of the issue at hand. Given the growing complexity of the negotiation process, proposing specific language on single issues is increasingly becoming a task for informal group Chairs, as subsidiary body Chairs, and even more so COP Presidents, have difficulty in keeping a detailed 'handle' on all aspects of the negotiations. In doing so, informal group Chairs will usually draw on advice from the secretariat.

COP Presidents and subsidiary body Chairs, however, often play a vital role in mixing and matching the proposals of parties to construct a whole acceptable to all. As a third party without a particular position to sustain, a presiding officer may be able to see the broad universe of possible combinations, not just the options in isolation, and then forge an integrative solution from these. A good example of this is the proposal by COP President Pronk made at COP 6 (part II), where he issued a paper – 'Core elements for the implementation of the Buenos Aires Plan of Action' (FCCC/Non-paper, 2001) – covering all the main issues under negotiation through a compromise mix of the positions of parties. Except for the section on compliance, which was subject to further negotiation, this formed the basis for eventual agreement (see also Chapter 11). Another example occurred on the final night of the Kyoto Protocol negotiations, when Chair Estrada correctly surmised that the combination of including emissions trading and the CDM in the Kyoto Protocol, but excluding voluntary commitments for developing countries, would prove acceptable to the parties as a whole. While an analysis of interventions made on that last night suggests that a majority might have supported the isolated decision to adopt an article on voluntary commitments, Estrada judged – probably correctly – that an alternative package including voluntary commitments but no emissions trading or no CDM would not have harnessed consensus.

Procedural innovation

Certain presiding officers have been particularly entrepreneurial in their management of the negotiation process, experimenting with new procedures and practices. In this sense, their contribution has been a more long term one, impacting on the overall procedural evolution of the climate change regime, and not just on the outcome of a specific negotiating round. Patrick Széll, for example, Chair of the AG13 from 1995–1998, pioneered an openness towards inputs from NGOs – both written and spoken – that is now being replicated in the permanent subsidiary bodies. Similarly, Chow Kok Kee, as SBSTA Chair, instituted regular informal meetings between the subsidiary body Chairs and NGOs which helped to foster a more inclusive model on NGO involvement, and also directly contributed to the active NGO involvement on the Expert Group on Technology Transfer. COP 6 President Pronk provides another example of a procedurally innovative presiding officer, in his case pioneering channels for stronger ministerial involvement (see Chapter 13). Although his success in doing so was mixed at COP 6 (part I), he did lay the foundations for more effective ministerial participation at COP 6 (part II) and COP 7.

Decision-taking

One of the formal functions of presiding officers, as set out in the rules, is to 'announce decisions'. This function acquires particular importance in the climate change context given the absence of an agreed voting rule. This, together with the lack of a clear definition of consensus, grants presiding officers some leeway in interpreting when decisions can be adopted (see also Chapter 8). So presiding officers are required to make a judgement as to whether any objections from parties are mere posturing or whether they reflect bottom line demands that should not be violated. Making a correct judgement in this regard is critical to the eventual legitimacy of the decision taken, as well as the presiding officer's credibility. As Kaufmann notes, 'all decisions taken without a vote require the careful judgement of the chairman. If his judgement ... errs, a procedural wrangle may well break out and the chairman's prestige will suffer' (Kaufmann, 1989, p30).

This decision-taking function is particularly important for COP Presidents and subsidiary body Chairs, given that formal action by the parties, whether adoption of decisions, conclusions, or any other type of output, can only be taken in plenary meetings of those bodies. Nevertheless, given the growing importance of contact groups and the tendency for the COP and subsidiary bodies to simply rubber stamp agreements reached in those contact groups, the role of contact group Chairs in registering a consensus that is then relayed to the main group is becoming more significant. Examples of such decision-taking are discussed in Chapter 8.

The decision-taking function of presiding officers means they are sometimes used as arbiters of last resort to decide between the various options on the table. In such cases, different parties and coalitions are unable to back down from their own preferred position and accept their opponent's proposal for fear of losing face, even if, on purely substantive grounds, they are prepared to do so. The presiding officer's intervention in deciding one way or the other can therefore be used as a 'lightning rod'. As noted above in the context of consensus building, parties can place responsibility – indeed, blame – for the decision on the presiding officer and therefore not appear to be backing down.

A good example is the approval of the article on general commitments for all parties in the Kyoto Protocol, which had been the subject of gruelling negotiations in an informal group in Kyoto. The Chair of the informal group, Bo Kjellén, reported privately to Estrada that there were three options on the table: one supported by Annex I parties, one by the G-77 and China, and a compromise option proposed by himself, but apparently acceptable only to Annex I parties. Estrada took several interventions on the draft article, which confirmed the various options, and finally asked the G-77 Chair what he should do, to which the delegate replied 'do what you always do, use your gavel' (CoW, 1997f). Estrada did precisely that, and Kjellén's proposed compromise was adopted without objection by either the G-77 Chair or any developing country; the G-77 Chair could not be seen to retreat after such intense negotiations, but he was prepared to do so in deference to Estrada's decision.

Inspiring and motivating

Providing inspiration and motivation to the negotiating parties to spur them on to agreement is an important leadership function that presiding officers can carry out. It is critical that presiding officers be committed to a positive outcome, albeit impartial as to the contents of that outcome and, equally critically, that they be convinced that agreement will be reached, even in the face of seemingly irreconcilable differences. Seeking to inspire and motivate parties has been a traditional role taken on by COP Presidents in the climate change regime, in line with their more ceremonial mantle.

COP 7 President El Yazghi is a good example of a President who sought to inspire and cajole negotiating parties to reaching agreement, often spilling over into stark warnings of the consequences of failure. He did so at a number of private meetings with representatives of the various negotiating coalitions and also in plenary, reminding delegates of the breakdown a year before at COP 6, and cautioning them of the consequences of failing again. Although the extent to which parties actually paid heed to the President as such is doubtful, the constant reminder of the failure at COP 6, and the President's unwavering conviction that success at Marrakesh was possible, did help to maintain momentum and a positive atmosphere.

Another example of the benefits of positive thinking is that of Chair Estrada, during the Kyoto Protocol negotiations. He was genuinely convinced of the reality of climate change, often referring in the AGBM to extreme weather events and their adverse impacts. He was therefore determined that the negotiations would be successfully concluded and maintained unfailing optimism in this. In his own words:

> You have to be optimistic. Particularly in Kyoto, I was paid to be optimistic...the Chairman has to be perseverant, persistent, otherwise you are lost...whatever happens you have to keep doing things up till you reach the goal (interview).

Such positive thinking was contagious; the fact that Estrada never admitted failure as an option injected energy into the negotiations and reinforced the unacceptability of failure.

An opposite example is that of COP 6 President Pronk in The Hague, who reportedly began expressing concern over possible failure by the close of the first week of negotiations. Although it was certainly realistic to examine fall back options, the absence of any clear conviction from the COP President that the meeting would succeed cast a palpable negative shadow over proceedings.

Interplay between presiding officers

COP Presidents, subsidiary body Chairs and informal group Chairs all have their own niches to fill and the interaction between these different layers of presiding officers can be as important to a negotiation process as individual performance. This is especially the case at COP sessions, where all three layers of presiding officers, including COP Presidents, are at work. Arguably the most successful COPs have been those where the COP President has confined him/herself to largely ceremonial duties, focusing leadership efforts on motivation, inspiration and broad-brush consensus building, including, importantly, setting deadlines, while allowing those with most experience in the negotiations, usually the subsidiary body Chairs and informal group Chairs, to concentrate unimpeded on the detailed negotiations. COP 7 is a good example of this; President El Yazghi gave a largely free rein to the (ministerial) Chairs he appointed to negotiate on his behalf, yet maintained pressure on parties through regular small group meetings. COP 3 is another good example, where the 'hands off' role assumed by the COP 3 President and the Japanese presidency more widely in the official negotiating arena was complimentary and beneficial to the more active leadership exercised by Estrada. The Japanese gave Estrada a free rein in the conduct of proceedings and organization of the negotiations, while focusing their efforts on forging deals through side payments and political pressure outside of the official negotiations. As one interviewee put it, 'it was Estrada that drove the process...and the fact that the Japanese let him do that was important'.

Skills and qualities

The above discussion has thrown light on a number of skills and qualities that can help enhance the performance of a presiding officer. Some traits are important prerequisites for any successful Chair, while others may be more or less appropriate for different types of negotiations and negotiating rounds. These issues are discussed further below.

Impartiality

As discussed above, impartiality is a necessary precondition for any effective presiding officer. Impartiality is distinct from neutrality. Presiding officers should not be neutral, in the sense that they should not be indifferent to the outcome of the negotiations, but instead be committed to reaching agreement. However, while not neutral, presiding officers should remain steadfastly *impartial* with regard to the differing preferences of parties and coalitions, as well as to specific issues.

Chair Estrada again provides a good example here. Estrada's determination to broker agreement during the Kyoto Protocol negotiations ran deep, and the presence of obstructionist parties meant that such determination was not compatible with neutral chairing. As one interviewee explained, 'he was totally committed to progressing negotiations towards a positive outcome, and...some of the parties weren't'. However, he did remain impartial on specific issues. Although, at times, Estrada made it clear that he was personally suspicious of certain proposals (e.g. emissions trading), he did not block the negotiation of any options based on his personal views if he thought they were necessary to get agreement.

If a presiding officer is perceived as biased, then his/her authority and therefore ability to guide the negotiations can suffer irretrievably. The prime example here is the Argentinian President of COP 4, who, in her opening statement, expressed support for a highly controversial proposal by Argentina to enable developing countries to take on voluntary commitments. Although she did not stand in the way of the removal of this item from the provisional agenda in the face of implacable opposition from most developing countries, the perception of her bias towards this item dogged her throughout the Argentinian Presidency.

Perceptions of bias can indeed be as important as actual bias. Although the

Dutch COP 6 President Pronk went out of his way to be sympathetic to negotiating coalitions other than his own, the G-77 still tended to interpret his actions negatively, while he also faced accusations of partiality from industrialized countries. The Umbrella Group perceived inherent bias towards the EU, whereas the EU detected compensatory swings towards the Umbrella Group, placing President Pronk in a very difficult position. Allegations of bias have also been levelled (albeit rarely openly) at Chairs of informal groups. These Chairs are often placed in an awkward situation, as members of their negotiating coalition may have expectations that their interests will be safeguarded, despite the need for the Chair to be impartial. Sometimes, the Chair him/herself may have a well-known position on the topic in question. Rumours of partiality surrounded the Chairs of the LULUCF, adverse effects, and technology contact groups at various points during the post-Kyoto negotiations, although these were not formally aired. Needless to say, in a highly contentious negotiating environment, it is almost impossible for Chairs always to foster the perception of neutrality. Interestingly, again illustrating the importance of experience and learning, the above-mentioned LULUCF contact group Chair, Halldor Thorgeirsson, then went on to a very successful tenure as SBSTA Chair, where impartiality was never in question.

Strength of personality

In all cases, a prerequisite for an effective presiding officer is a strong personality that can withstand pressure and is not afraid to take decisions in the face of conflicting views. This is especially the case in the complex and conflictual climate change negotiating environment. An effective presiding officer must be able to maintain focus and direction in order to sustain the efficiency of the negotiations, and not be side tracked or cowed into changing course by strong willed delegates. Chair Chow, for example, in his chairing of the post-Kyoto contact group on flexibility mechanisms, was subject to very great pressure from delegates to modify his approach to the negotiations, but largely succeeded in holding his own ground. Chair Thorgeirsson was particularly praised by interviewees for being unafraid to tackle difficult issues in the negotiations head on, rather than being content with an 'agree to disagree' result.

Chair Estrada was perhaps the most famously forceful presiding officer, who stamped his authority on the Kyoto negotiations. At an AGBM session in early 1997, for example, he sought to curb the lengthy restatement of positions by only allowing interventions from parties making concrete proposals. To enforce this ruling, Estrada interrupted several delegates in the midst of their interventions, a highly unusual practice in UN forums (see Chapter 7). Having done so, he declared, 'I apologise to those I interrupted... Next time, there will be more! I hope you understand why' (AGBM 6, 1997). Estrada exerted his authority to push negotiators into a constructive negotiating mode, challenging any party whose stance he thought was uncooperative, particularly the Annex I parties. He frequently chided Japan and the US, for example, for their tardiness in putting forward proposed emission targets, while also reproaching the EU for its cumbersome internal decision-making process. While the actual impact of such rebukes is difficult to quantify, they did maintain constant pressure on the Annex I parties. As the negotiations advanced, Estrada grew bolder in also challenging the G-77.

Estrada's forcefulness, however, meant that, in the words of a secretariat interviewee, he sometimes 'bruised people'; his aggressiveness could be counterproductive and cause offence. In Kyoto, for example, he accused the Brazilian delegation of having come to the conference 'with an open hand' (CoW, 1997c), a reference to Brazil's proposed clean development fund. The suggestion that the fund had been proposed for Brazil's financial gain triggered a walk out by the Brazilian Ambassador. A couple of interviewees remarked that Estrada was not universally liked and had annoyed several delegations. However, it was critical to the process that Estrada was not afraid of unpopularity. As one NGO interviewee commented:

> Having someone with his personality and his initiative to...take control...to be unpopular at times, and not to please absolutely everyone 100 per cent of the time, is necessary in a strong Chairman, otherwise you'll never get an agreement...That he was willing to... take on that role was absolutely essential.

However, in an instructive illustration of how differing chairing styles may be more or less appropriate for different negotiating phases, Estrada's forcefulness was less appreciated, and indeed less effective, in his chairing of the SBI in the post-Marrakesh period.

Experience

Experience of complex global negotiating arenas, especially on the specific issue of climate change, can be of immense importance to a presiding officer's performance. Indeed, most subsidiary body Chairs tend to be appointed from among experienced delegates, having acquired the network of support needed to be nominated by their regional group. This is not the case, however, for COP Presidents, who often assume their post by virtue of having offered to host the COP. Even when this is not the case, COP Presidents rarely have experience of chairing global negotiations, and have often not even attended a climate change meeting beforehand for more than a few hours.

The importance of experience is highlighted by the fact that several subsidiary body Chairs have improved their performance during their tenure. Thorgeirsson and Chow were both cited in this respect by interviewees. Several noted how Thorgeirsson had 'grown into his role', growing more confident and focused after a hesitant start. Similarly, after a shaky initial performance, Chow developed his own chairing style as SBSTA Chair that harnessed sufficient confidence among the parties for him to be accepted as sole Chair of the flexibility mechanisms contact group, a pivotal position in the post-Kyoto negotiations.

Unfortunately, COP Presidents rarely have the chance to learn from experience and improve their presiding skills, as they are only in place for one year, and usually only for one negotiating session. The example of COP 6 President Pronk is again instructive. He was able to learn from the difficulties he faced at COP 6 to develop a much more effective presiding style eight months later at COP 6 (part II), where his understanding of the political dynamics of the climate change negotiations and how to manage these improved dramatically, and his less dogmatic, yet at the same time more confident, approach were critical to achieving a consensus.

Technical knowledge

Interviewees agreed that technical knowledge and political understanding of the issues at hand are important for a presiding officer. Without such knowledge and understanding, it is impossible to take informed and judicious decisions, let alone present substantive proposals. While understanding of the politics involved in an issue is always necessary, the importance of technical knowledge varies depending on the level of presiding officer. It can be very important for informal group Chairs, but is obviously much less so for COP Presidents who, as ministers, cannot be expected to have a deep appreciation of the technical issues involved. Indeed, according to one interviewee:

> one of the greatest formulas for disaster in a multilateral negotiation is having a person who's a super technician in charge... because the use of those skills has all the makings of creating friction... perhaps being too possessive of ideas and therefore inflexible, because after all, you know the subject so well, you know you are right.

Technical knowledge has, however, proved very important to the effectiveness of SBSTA Chairs. Thorgeirsson and Harald Dovland, for example, were both technically minded Chairs who used their knowledge to guide the negotiations to good effect. Similarly, SBSTA Chair Chow had a sound understanding of issues relating to technology transfer, and was able to establish a lasting positive legacy in the climate change regime on that topic.

English language skills

A good, but not necessarily perfect, command of the English language is necessary to preside over a global negotiation. Despite the availability of simultaneous interpretation and document translation, negotiations are still almost always conducted in English, especially in the final stages. Even in plenary settings, inability to chair in fluent English can be a real obstacle. SBI Chair Bakary Kanté and COP 7 President El Yazghi both experienced these obstacles. Kanté tried to chair in English, but faced difficulties in being understood and authoritative. He was more effective when speaking in French, but some of his more subtle points were inevitably lost in translation. COP 7 President El Yazghi, for his part, was hindered in his attempts at consensus building by the fact that he could not speak directly to Anglophone heads of delegation, including the English-speaking Iranian G-77 Chair, and had to work through an interpreter.

Native English, however, is not necessary, and indeed imperfect English can benefit a presiding officer. 'English is not my mother tongue' is an excuse commonly used by delegates, as well as presiding officers, to question a controversial proposal or explain away an unpopular suggestion. Imperfect English can indeed allow a presiding officer to speak in a more direct and candid manner than would be diplomatically acceptable for a native Anglophone, a trick that Estrada often used to good effect. Imperfect English can also be used by a presiding officer to generate affinity with the majority of parties, which are not native English speakers. Estrada, for example, although he chaired in English, addressed his fellow Hispanic delegates in Spanish, thereby underscoring his empathy for non-Anglophone developing countries, and the added difficulties these face in participating in the negotiations.

Nationality

The impact of nationality on effective chairing is difficult to gauge with confidence. In general, developing country presiding officers can enjoy a degree of goodwill from the largest coalition, the G-77, that an Annex I party chair would find difficult to acquire. Furthermore, developing country presiding officers tend to have more latitude in their dealings with other developing countries, being able, according to one interviewee, to 'speak and address both developed and developing countries'. According to a European interviewee, 'the developing countries can be rude to us and each other, but we [the Annex I parties] can't be rude to them'. However, a number of very effective presiding officers, including Thorgeirsson, Dovland and Széll, have all come from industrialized countries, suggesting that other skills and attributes are more important than nationality.

Diplomacy and energy

Diplomacy is, of course, very important for the presiding officers of global negotiations, where a diplomatic faux pas can lead to loss of confidence and respect. An interviewee highlighted an incident whereby a subsidiary body Chair had committed the 'diplomatic faux pas of interrupting China', which he claimed had adversely affected relations between that Chair and developing country parties. Linked to the need for diplomacy is infinite patience – as one interviewee put it, 'frustration must not be allowed to intrude'. It is certainly very hard to maintain diplomacy in the face of so many hotly competing points of view and approaches, covering both sincere concerns and deliberate obstructionism. Even Estrada, in the final days of the Kyoto negotiations, found that his natural forcefulness was losing its diplomatic edge and turning into counterproductive aggressiveness.

A host of other personal qualities are needed to effectively preside over a global negotiation. A large reserve of energy is one of these. Negotiations often continue round the clock, and the presiding officer must be able to sustain the vigour needed to make judicious decisions. Estrada was notable in this respect. One interviewee commented, 'I don't know where he got his energy from, I really don't. But to keep the momentum is the thing that solved Kyoto... he kept people off balance by his energy in keeping the whole thing going.' Estrada's energy was due not only to his character, but also to good personal management. He attached great importance to his rest and nutrition so that he had energy in reserve when it mattered.

Summary and concluding remarks

This chapter has identified certain qualities that can contribute to the success of presiding officers, notably diplomacy, patience, energy, chairing skills, English language skills, impartiality and strength of personality. One thing is clear, however – there is no such thing as a set template for a successful presiding officer. A one size fits all approach to leadership is not helpful. Different types of negotiation need different types of leadership at different stages in the regime building, development and strengthening processes. Different bodies also need different leaders. The attributes required of a presiding officer who must regularly chair full plenary meetings of 185 countries are very different from those required of a presiding officer whose main mandate is to 'wheel and deal' a consensus among a few key players.

Two basic related principles needed for effective leadership stand out – continuity and experience. As discussed above, the Kyoto Protocol negotiations benefited from having a single leader throughout the negotiation process, who was able to gain the confidence of parties and establish his authority over time, serve as a focus for the negotiations, and strategically plan the process, while the post-Kyoto negotiations suffered from the absence of any continuity in leadership. In addition, the case of the climate change regime suggests that, to be effective, presiding officers should have experience of the relevant regime, its political complexities and sensitivities. In a number of cases, presiding officers have also improved their performance considerably, as they have built up experience.

However, with the exception of the critical first factor relating to a single continuous leader, the extent to which such factors, including experience, can actually be implemented in the selection of a presiding officer is restricted, given that the election of officers is usually done on the basis of political deals between and among regional groups without, for example, the involvement of the secretariat or another objective party with the overall interests of the process at heart. Although it would be wrong to say that personal qualities and qualifications are not taken into account in the election of presiding officers, they can play second fiddle to regional politics. Unfortunately, therefore, despite its central importance in the organization of a negotiation process, the presence of an effective presiding officer possessing the right background and experience – whether for the COP, subsidiary bodies, or informal groups – can rarely be assured.

Bureaux

The Bureau is necessary... but it doesn't always work (interview).

Introduction

This short chapter discusses the role of the COP Bureau in organizing the climate change negotiation process. In doing so, it also touches on the roles of the other Bureaux in the regime, namely the Bureaux of the two permanent subsidiary bodies – the SBSTA and SBI – and the Bureaux of the two former ad hoc bodies, the AGBM and AG13.[1] The chapter begins by looking at the formal mandate and functions of Bureaux, before examining the actual contributions that Bureaux have made to the climate change regime at different points in the negotiation process.

Composition

COP Bureau

It is common in multilateral negotiations for the presiding officer to be assisted by a Bureau, composed of delegates elected by the parties. As set out in the rules of procedure, the COP Bureau consists of 11 officers.[2] In addition to the COP President and Chairs of the two permanent subsidiary bodies (see Chapter 4), these officers comprise seven Vice-Presidents and a Rapporteur. Given that the positions of COP President and subsidiary body Chairs are discussed in detail in Chapter 4, this chapter focuses mostly on the roles of the Vice-Presidents and Rapporteur.

The rules of procedure specify that each of the five UN regional groups – Africa, Asia, Central and Eastern Europe, Latin America and the Caribbean and Western Europe and Others – must be represented on the Bureau by two officers. The eleventh post is reserved for the small island developing states, given their particularly great stake in the climate change regime. The establishment of this dedicated seat under the rules of procedure was disputed at COP 1, with some developing country oil exporters (notably Saudi Arabia and Kuwait) arguing that they too should hold a special seat on the Bureau, on the grounds of their potential economic vulnerability to the impacts of response measures. Although no formal rule was agreed to this effect, the demands of the oil exporters have been accommodated through an informal understanding that they will be represented through one of the regional groups. In practice, this has meant that OPEC has been represented on every COP Bureau since COP 2. Bureau members from eligible countries receive funding to cover the costs of their attendance at meetings, in addition to the delegate(s) already funded from their country through the Trust Fund for Participation (see Chapter 7).

As discussed in Chapter 4, Bureau members are elected through nominations by the regional groups, which are then rubber stamped in the COP plenary. Membership of the COP Bureau is highly sought after, as it is perceived as the inner circle of the regime, granting access to information and influence (Yamin and Depledge, 2004). It can prove difficult, however, to fill the post of Rapporteur, as this is considered a junior post in the wider UN system. Regional groups – especially the large, heterogeneous ones, such as Asia – sometimes find it difficult to elect their representatives. At COP 9, for example, it proved impossible to elect the Bureau until the last day. In accordance with the rules of procedure, the outgoing Bureau thus continued to meet to fulfil the Bureau functions. Officers may be re-elected, but may not serve for more than two terms.[3]

Subsidiary body Bureaux

In addition to the Chair, the subsidiary body Bureaux consist of only two members, a Vice-President and a Rapporteur. There is no formal requirement for regional rotation, but achieving an acceptable geographical distribution over time is an important consideration.

The rules of procedure specify that subsidiary bodies other than the SBSTA and SBI should also have three-person Bureaux, elected with 'due regard to the principle of equitable geographical representation' (Rule 27). This approach was indeed followed for the AG13. The political significance of the AGBM, however, and the perception that membership of its Bureau would wield valuable power and influence at least over procedural aspects of the negotiations, meant that a three-person Bureau proved impossible to agree, largely due to OPEC's insistence on having its own seat.

In an excellent example of the importance of pragmatism and improvisation in the interpretation of the rules of procedure, an innovative solution was agreed at the third session of the AGBM in 1996. This provided for the formal three-person Bureau to be supplemented by six 'advisers', plus the Chairmen of the SBSTA and SBI ex officio. The total number of de facto Bureau members thus rose to 11, allowing the agreed COP formula to be applied to the AGBM, including representation of OPEC through one of the regional groups. The solution also provided for the post of Rapporteur to rotate between the two full Bureau members who would both be known as Vice-Chairmen. Giving both full Bureau members the same status thus removed an area of competition between the regional groups.

Mandate and functions

The only tasks of the COP Bureau (except for the President) that are specified in the rules of procedure are to examine the credentials of representatives to negotiating sessions and submit a report on these to the COP, and for the Vice-Presidents to replace the President in case of absence (Rules 20 and 24).[4] The Vice-Presidents are usually called upon in the latter case for the chairing of the high-level general debate (see Chapter 13). The Rapporteur has formal responsibility for the production of the COP report, which is presented in his/her name. However, his/her role usually amounts only to a brief overview and rubber stamping of the secretariat's work (Yamin and Depledge, 2004).

The Bureau has assumed a couple of other procedural functions, either through direct mandate by COP decisions, or unwritten practices that have introduced tasks for it. Pursuant to decision 14/CP.1, the UN Secretary-General must consult with the COP through the Bureau before appointing the Executive Secretary of the Secretariat. The extent of the Bureau's involvement in this regard was a bone of contention among Bureau members during the appointment of the second Executive Secretary, Joke Waller Hunter. The implications of this controversy and insights that can be derived therefrom are discussed below. In terms of unwritten practices, the Bureau reviews the list of new NGOs/IGOs requesting admission, once these have successfully passed initial clearance by the secretariat and before they are presented to the COP (see Chapter 14). This is an important role prior to subsidiary body sessions, as applicant organizations are admitted as observers to those sessions on the basis of Bureau scrutiny, pending formal action by the COP at its following session. However, although procedurally important, in practice the scrutiny of applicant organizations is almost always a formality and serious questions are rarely raised.

Like the role of the presiding officer, however, lack of detail in the rules of procedure belies a much greater informal role for the Bureau in providing process-oriented leadership and organizational energy. The informal role taken on by the Bureau in this regard varies from session to session, depending on the personalities and attributes of its members, the preferred approach of the President, and the needs of the negotiation process. A key underlying factor to the work of the Bureau is that, unlike the President, Bureau members *represent their constituencies* – that is, the nominating regional groups – and are not, therefore, required to be neutral. The subsidiary body Chairs must toe a particularly delicate line when serving on the COP Bureau, being required to maintain the impartiality that befits a presiding officer, yet also being required to represent their constituencies. COP Bureau members are expected to elicit the views of their constituencies, and report back to them. However, they are also expected, as individuals, to promote the interests of the process, and therefore to supply constructive, rather than confrontational, input to the negotiations.

The composition of the Bureau – based on the formal regional groups rather than the negotiating coalitions – means that it is used almost exclusively for procedural and organizational matters, rather than substantive negotiation (Yamin and Depledge, 2004). Indeed, Bureau members have no mandate from their constituencies to deal with substantive issues. For this reason, at highly political COP sessions, the COP Bureau membership is often supplemented by representatives of negotiating coalitions, to form an 'Expanded Bureau'. The Expanded Bureau is then in a better position to take on substantive issues, as well as procedural issues with a substantive component to them. The role of Expanded Bureaux is discussed in more detail in Chapter 9.

Unlike the position of President, the other Bureau posts are served at the level of officials. The exception is for meetings during the high-level segment of COP sessions, where countries on the Bureau will often be represented by attending ministers.

The functions of the subsidiary body Bureaux mirror those of their counterparts on the COP Bureau. The subsidiary body Vice-Chair is expected to replace the subsidiary body Chair in the event of absence, while the subsidiary body Rapporteur is formally responsible for the subsidiary body report. The subsidiary body Bureaux, however, have no other specific tasks akin to the appointment of the Executive Secretary or scrutiny of applicant NGOs.

Informal roles

COP Bureau members are expected to supply process-oriented leadership and organizational energy to the negotiation process in a number of ways. One of these is by chairing contact groups (Yamin and Depledge, 2004). COP Bureau members have been called upon to (co)chair contact groups in the hope that this will be a non-controversial, accepted appointment. Members of the COP Bureau were especially drawn upon in this way at COP 6 (part II), following some controversy over the chairing of informal groups. There is no guarantee, however, that COP Bureau members will possess the necessary skills to chair contact groups, so the extent to which they can be drawn upon is, in practice, limited. In a similar vein, COP Bureau members are routinely called upon to consult informally on sensitive or procedural issues, such as the election of the forthcoming Bureau.

Perhaps the most important informal role of the COP Bureau is to serve as a consultation forum, or early warning system, for the COP President (see also Yamin and Depledge, 2004). As noted in Chapter 4, COP Presidents are typically ministers with limited experience of the climate change process. The other COP Bureau members, who are usually officials with experience of the climate change negotiations, can therefore provide important informal guidance and advice to the COP President on how best to achieve a successful outcome. To reiterate, however, this will almost always concern procedural and organizational, rather than substantive, matters. The COP President, for example, may ask for advice on how to handle a contentious agenda item, or on how to organize the final stages of negotiations (e.g. whether and how to convene a friends of the chair group). Bureau members, for their part, may consult with their constituencies, and provide initial feedback on the acceptability of various proposals. It can be very important to seek feedback on potential approaches in this way, before formally proposing them, so that unanticipated objections or problems can be overcome. Similarly, Bureau members themselves may spontaneously raise issues or problems of concern to their constituencies in COP Bureau meetings, thereby providing advance indication of underlying tensions in the negotiations. As one interviewee put it, the Bureau acts as a 'mini plenary' or 'focus group', providing a foretaste 'of how things will play out'.

A good example here concerns the AGBM Bureau. At an AGBM Bureau meeting in 1996, Saudi Arabia, a Bureau member, raised questions regarding the six-month rule, requiring the draft text of a protocol under negotiation to be circulated six months before its adoption. This prompted Estrada to seek a legal opinion from the UN Office of Legal Affairs, which was pivotal in deflecting the possibility that the six-month rule might be interpreted in a problematic light (see Chapter 7). Estrada also used the AGBM Bureau to convey messages to the regional groups, which he hoped would promote progress at the session. For example, immediately prior to AGBM 8, the last session before COP 3, he asked the Bureau to make it clear to their regional groups that he would not allow square brackets (in UN and similar negotiating arenas, language that is not agreed appears in square brackets) to be reinserted into the Chair's Text (AGBM Bureau, 1997).

The Bureau is also called upon to supply organizational energy in responding to, and providing solutions for, organizational and procedural issues and problems that may arise in the negotiation process. To this end, during the finales of the post-Kyoto negotiations at COP 6 (parts I and II) and COP 7, Chairs of the key informal groups were asked to report to COP Bureau meetings on progress in their negotiations. At COP 7, for example, the Bureau arbitrated on problematic timetabling issues for contact groups, decided whether negotiating groups should be open or closed to observers, and set deadlines for work. The Bureau's role in taking such procedural decisions seems to be increasing. Where new procedural questions have arisen requiring some form of formal sanction, the COP Bureau has been called upon as a decision-maker on behalf of the regime. A good example here concerns requests by indigenous peoples and research and independent NGOs to form new constituencies, which, despite the absence of any rule stating that this was necessary, were considered and sanctioned in the COP Bureau (see also Chapter 14). This reflects the growing complexity of the climate change regime, whereby the secretariat hesitates to take action on significant procedural decisions on its own authority, yet taking such decisions to the COP or subsidiary body plenaries would be an inefficient use of resources. The COP Bureau thus serves as a useful intermediary body.

The Bureau's decisions, however, are not officially circulated, and are often just communicated orally by members to their constituencies, or by the President or other presiding officers to the relevant negotiating forum (Yamin and Depledge, 2004). This has, on occasion, caused controversy, as the recollection of decisions by different Bureau members has varied. Despite this, Bureau decisions are typically well respected, and the authority of the Bureau is often invoked by presiding officers to support their own actions. In this sense, the Bureau can act as a legitimating forum. Presiding officers can justify decisions on the organization of the process – allocation of interpretation, scheduling, deadlines, use of negotiating texts – by saying that these have been cleared by the Bureau.

This legitimating function is particularly useful when decisions are needed outside the official negotiating sessions. The Bureau then serves as a legitimate forum for taking procedural decisions that are too great for the secretariat to take by itself. The best example here concerns the date and venue of the forthcoming COP session, where information is often insufficient at the previous COP session, and the Bureau is therefore asked to take a decision inter-sessionally based on, for instance, a secretariat technical mission. This occurred, for example, at COP 9, when the Bureau was asked to consider further information to be provided by the bidding city for COP 10, Buenos Aires. Another case is the organization of the high level ministerial segment at a COP session. At SBI 18, for example, the SBI decided that roundtable discussions would be held at the forthcoming COP 9 but asked the COP Bureau to 'give further consideration to the details and format' of these (e.g. see SBI 18 report, paragraph 43e). In response to this mandate, the COP Bureau decided on the number, themes, timing and participation at the roundtables (e.g. see FCCC/CP/2003/1/Add.1), with its decisions subsequently endorsed by the COP at its opening plenary.

It is perhaps inevitable that, in some cases, individuals have used the opportunities afforded to them by their membership of the Bureau and exerted organizational energy to pursue politically motivated topics or issues of personal concern. An example here is the way in which the procedure for the appointment of the new Executive Secretary (upon the retirement of the first Executive Secretary, Michael Zammit Cutajar) was questioned by a Bureau member, who, it is alleged, was motivated by the desire of his constituency to veto the appointment of a particular candidate.

In some rare instances, the Bureau has exerted real process-oriented *leadership*; here, leadership is distinguished from the more routine supply of organizational energy, in the sense that the Bureau has, through an unsolicited initiative, filled a vacuum in direction to advance the climate change process. The best example of this is from COP 6 where, following the collapse of negotiations, the Bureau met and urged COP President Pronk to suspend the session, and reconvene it a few months later. This procedural solution, which was proposed by individual members and supported by the Bureau as a whole, helped to avoid declaring the session a failure, while being organizationally simpler than the convening of an extraordinary COP. Here, the COP Bureau was able to step in and take over at a time when the COP President and parties were exhausted and unable to come up with a procedural solution to sustain the momentum of the process. Individual members have also taken the lead at particular points in time to provide sorely needed direction to the process, although there are few such instances. At COP 6 (part II), for example, Vice-President Raúl Estrada helped move negotiations forward by offering to prepare a single document bringing together the outputs of the individual negotiating groups as input to the high-level segment. This document (FCCC/CP/2001/CRP.8), clearly listing issues and options, proved to be very useful in structuring the subsequent negotiations (see also Chapter 11).

The Vice-Chairs and Rapporteurs on the subsidiary body Bureaux may perform similar roles – acting as an early warning system or focus group, legitimizing decisions, taking decisions in between sessions, or taking the lead where necessary. However, given the more technical and lower profile nature of the work of the subsidiary bodies, it is rare that subsidiary body Bureau members have much scope to exercise these informal functions. For the most part, subsidiary body Bureau members do not take on much of a role in supplying organizational input, energy or leadership, but rather stick to their (very few) formal functions.

Skills and qualities

The activities of each COP Bureau and its performance will depend on two main factors: the use that the President wishes to make of the Bureau, and the skills and qualities of its members. Regarding the former, most COP Presidents have welcomed the potential contribution of the Bureau, although some have been more proactive than others in seeking input and advice.

In terms of the skills and qualities needed in a Bureau, perhaps the most important is experience. Bureau members with longstanding knowledge and understanding of the climate change negotiation process can be immensely helpful to a COP President. The fact that John Ashe and Raúl Estrada, two highly experienced delegates, were on the Bureau at COP 6 certainly helped the Bureau to take the lead and offer the procedural solution to the collapse of negotiations discussed above. Unfortunately, however, membership of the COP Bureau can be seen as a perk, given that it confers prestige, as well as additional financial support for delegates to attend negotiating sessions. Some constituencies will therefore nominate representatives who do not have experience of the negotiations, and who therefore find it hard to contribute constructively. Almost every Bureau will therefore have its silent members who rarely speak.

An ability to present a balanced point of view and constructive suggestions, while at the same time representing a constituency, is also important. The greatest danger within a Bureau is indeed politicization, where some Bureau members do not act, or are perceived as not acting, in the interests of the process, but exclusively to further their constituency's – or personal – positions. This was the situation for the AGBM Bureau, where wrangling over its membership meant that, once appointed, it was seen as a highly political body, some of whose members were known as obstructionists. The politicization of the AGBM Bureau meant that less use was made of it by the Chair than he might otherwise have done.

Summary and concluding remarks

The informal roles taken on by Bureaux have varied from session to session, as have their performance. Most interviewees, whether interviewed in 2000 or 2003, were rather lukewarm about the respective COP Bureaux they had experienced. One interviewee, who had served on the COP Bureau, commented, 'it's so-so. It depends a lot on the people on it.' Others stated that representation was 'not particularly star quality' and the Bureau's role 'not terribly impressive'. This impression is partly due to the silent members on the Bureau who rarely make a contribution, and partly to the obvious political agenda of some others.

Nevertheless, the COP Bureau has grown in importance over time, as the complexity of the climate change negotiation process has increased, and more organizational decisions with political implications are needed. In some cases, it is becoming increasingly difficult for the secretariat or presiding officers alone to take a decision – on the scheduling of contact groups, say, or allocation of interpretation – so that the Bureau is called upon to arbitrate. The Bureau also appears to be taking more decisions inter-sessionally. At present, however, the role and functions of the COP Bureau remain ad hoc and unwritten, and still vary greatly depending on the needs of the negotiation process and the approach of the COP President. For most delegates, the COP Bureau remains a rather shadowy body whose nature is poorly understood. As the climate change regime becomes organizationally more complex, it is likely that the role of the Bureau will continue to expand. This could imply the need for more formalization of its role and more transparency in its dealings, including, for example, the publication of its decisions and reports on its meetings.

The Secretariat

The secretariat can't deliver success, but it can deliver failure (secretariat interview).

Introduction

The secretariat is usually the only full-time actor within a regime, remaining active between sessions of the COP and subsidiary bodies. Like presiding officers and bureaux, secretariats of different regimes, and at different points in time, can vary greatly in their levels of activism. Wettestad (1999), for example, distinguishes between 'assistant' and 'player' secretariats, Sandford (1994) makes a distinction between 'actors' and 'stagehands', and Miles et al (2002) identify a range, from secretariats 'confined to office and record-keeping functions' to those providing 'political inputs' and 'promoting own ideas and solutions'. The importance of a competent secretariat – whether active or passive – to the success of a negotiation process is recognized in the literature, although it is not seen as pivotal, and certainly less important than that of the presiding officer (see Lang, 1989a; Sandford, 1994; Andresen and Skjaerseth, 1999; Wettestad, 1999; Miles et al, 2002). However, this more modest role attributed to secretariats may be a reflection of the fact that secretariats prefer their activities not to be noticed; as we shall see in this chapter, invisibility may thus be a sign of effectiveness, rather than irrelevance.

This chapter explores the role of the secretariat in the climate change negotiations, and how it has changed over time in tandem with each successive negotiating phase. In doing so, it focuses almost exclusively on the secretariat's role in providing support to the negotiation process, rather than its programmatic activities.[1]

Institutional arrangements

In the same resolution that launched negotiations on the Convention in 1990, the UNGA requested the UN Secretary-General, in consultation with the heads of UNEP and the World Meteorological Organization (WMO), to establish an 'ad hoc' secretariat to provide support to those negotiations (UNGA Resolution 45/212, paragraph 12). As is commonplace in international treaties, the Convention itself then called on COP 1 to 'designate a permanent secretariat and make arrangements for its functioning' (Article 8.3). COP 1 duly did so, in effect confirming the existing Intergovernmental Negotiating Committee (INC) secretariat as the permanent Convention secretariat (decision 14/CP.1). Also at COP 1, parties decided (through a secret ballot, the only 'vote' ever taken in the climate change regime) to accept an offer from Germany to host the secretariat headquarters (which had, up to then, been located in Geneva) (decision 16/CP.1). The permanent secretariat moved to Bonn in August 1996 (Yamin and Depledge, 2004).

The secretariat is 'institutionally linked to the United Nations' (decision 14/CP.1), and thus administered under UN rules and regulations – for example, concerning staffing structure, finance and administration. The core funding for the secretariat is secured through financial contributions from all parties, their shares being based on the UN scale of assessment. In addition to the core budget, other trust funds administer voluntary contributions to cover the costs of participation at negotiating sessions of eligible delegates from developing countries and EITs (see Chapter 7), and to finance specific additional projects and activities. The level of the core budget, and requested levels of the other funds, are set in the programme budget, which is adopted by the COP every two years. The core budget of the climate change regime has almost doubled since entry into force of the Convention, from just over US$9.2 million for 1996 to just over US$17.3 million for 2005 (see decisions 17/CP.1 and 16/CP.9). Although it is now slowing down, this rapid growth bucks the general trend in UN institutions, and reflects the importance attached by parties to the issue of climate change, along with the rapid development of the regime's rules, including the adoption of the Kyoto Protocol (Yamin and Depledge, 2004).

The secretariat is staffed by UN civil servants. The head of the secretariat, the *Executive Secretary*, is appointed by the UN Secretary-General (in consultation with the COP Bureau – see Chapter 5). Since 2001, the Executive Secretary has been assisted by a Deputy. Next in the hierarchy are the handful of *programme coordinators*; of particular consequence for this book are the coordinators responsible for each of the three permanent bodies: the COP, SBSTA and SBI. Then come what might be termed the *issue managers*, who manage the secretariat's work on particular topics, technology transfer, GHG inventories, public outreach and so on. Working under the issue managers are the remainder of the lower-ranking professional – known as 'P' – staff, along with the general – or 'G' – staff, making up the secretarial, administrative and clerical workforce.

A striking feature of the secretariat in organizational terms is how rapidly its staff has grown, indeed faster than the increase in the core budget. In January 1995, just before COP 1, the secretariat consisted of 34 staff (20 P). Eight years later, in 2003, the number of approved posts in the secretariat had risen nearly five-fold to 168.5 (93 P).[2] The very rapid growth of the secretariat has inevitably brought with it some organizational challenges. These were most in evidence during and immediately after the Kyoto Protocol negotiations, that is, following the move of the headquarters from Geneva to Bonn. The secretariat is now a much more settled and stable organization. Throughout its history, and despite the upheaval associated with rapid growth and the move to Bonn, the secretariat has benefited considerably from the continuity in its senior staff, notably the first Executive Secretary, who served for over a decade.

Mandate and functions

The mandate and functions of the secretariat are circumscribed by four main sources:

- the Convention
- the rules of procedure
- the biennial programme budgets and
- specific requests made by the COP and subsidiary bodies.

The Convention sets out the general functions of the secretariat in a 'rather standard way for international environmental politics' (Wettestad, 1999, p217), and these functions are elaborated on a little in the rules of procedure. In terms of support to the negotiations, both the Convention and the rules of procedure emphasise the *logistical* role of the secretariat. 'To make arrangements for sessions' of the COP and subsidiary bodies and 'to provide them with services as required' is the first function set out in the Convention. The rules of procedure, in turn, specify particular functions in relation to interpretation, documentation and record-keeping.

The programme budget sets out a more detailed work programme for the secretariat for the coming two years. The COP and subsidiary bodies then request specific tasks, such as preparing documents, organizing workshops, compiling inputs from parties and disseminating information. While the secretariat is bound to comply with the requests of the regime bodies, it has, when necessary, made it clear that insufficient resources are available, or a request is simply impractical or procedurally inappropriate.

The 'Staff Vision' of the secretariat (see www.unfccc.int), prepared by secretariat staff members themselves, provides a succinct and useful summary of the secretariat's role by identifying three key tasks:

- providing organizational support
- supplying technical expertise and
- facilitating the flow of authoritative information.

Just like the presiding officers discussed in Chapter 4, the rather loose definition of the secretariat's functions means that it has the potential to take on a much broader informal role in helping to steer the negotiations to success through the supply of organizational energy, or even process-oriented leadership. However, the extent to which the secretariat can exercise this potential for activism rests on the delicate balancing of the differing pressures and expectations upon which its role is constituted.

The secretariat's two masters

Similar to the presiding officers, the secretariat is expected to help steer the negotiations to a successful conclusion. However, this expectation does not extend to one of true *leadership*; the parties do not expect the secretariat to lead, but rather to assist them. In the minds of parties, the secretariat is their servant, not their leader, and this also constitutes the public view of the secretariat. The secretariat Staff Vision, for example, prefaces the statement of its role by saying that it is 'guided by the parties to the Convention'. At the same time, however, the permanent organizational structure of the secretariat can generate tension with parties, most of which guard their sovereignty jealously and are sensitive about relinquishing powers to an intergovernmental organization. The secretariat must therefore ensure that it can justify its actions against its mandate and that it manoeuvres within its space for action even more carefully than the presiding officers who are not seen as a potential threat to sovereignty.

The most important commodity for a secretariat – more than efficiency or even competence – is *impartiality*, and confidence in that impartiality. As a body of international civil servants, the secretariat is required by UN Staff Rules and Regulations to serve the parties in an unbiased manner and not to promote or support any particular point of view (see ST/SGB/2002/1, regulation 1.2; Yamin and Depledge, 2004). This is even more important to the successful development of the regime than the necessary impartiality of presiding officers. While any suspicion of bias in a presiding officer would only affect the regime during the individual's term of office, perceptions of partiality within the secretariat would be a much more persistent problem that could put the whole process in jeopardy.

However, while the secretariat is impartial towards the various contrasting positions of the parties, it is not *neutral*, in the sense that, as a body, it is dedicated to promoting the success of the regime (Yamin and Depledge, 2004). The secretariat Staff Vision emphasizes from the outset that the secretariat supports 'cooperative action by states to combat climate change and its impacts on humanity and ecosystems'. It therefore takes as given that climate change is a problem, and one that necessitates a response centred on international collaboration. The difference between impartiality and neutrality was explained thus by former Executive Secretary Zammit Cutajar in his interview:

> The secretariat is objective [impartial], but not neutral. I say this because of a wise saying of Raúl Prebisch, my first boss in [the UN Conference on Trade And Development] UNCTAD. He said 'as a secretariat we are objective, but we cannot be indifferent to development. We cannot be neutral. We are fighting for development'. So when people try to block the [climate change] process, we can admire their negotiating skill, but we cannot be indifferent. We are here for a reason, not just a take home salary. We have a commitment.

Fundamentally, the secretariat is thus required to serve two masters: on the one hand, as the guardian of the climate change regime, it is obliged to support the ultimate objective of that regime and therefore both to address climate change in a meaningful way, and to achieve success in the ongoing negotiations. On the other hand, the secretariat is subservient to, and dependent on, the will of the parties, having been established to serve them (Yamin and Depledge, 2004). Given the presence of obstructionist parties in the regime, these two goals do not always coincide. As Sandford notes, 'secretariat objectives may be at odds with the objectives of individual treaty parties, even though secretariat objectives may be aligned with those of the treaty itself' (Sandford, 1994, p23). This tension has always been inherent in the secretariat's work, and has occasionally surfaced in dealing with obstructionist parties, as alluded to by the Executive Secretary above and as discussed further below.

The tension between the secretariat's two masters recently came to the fore in the context of uncertainty over the Kyoto Protocol's entry into force, and the fact that some powerful parties (e.g. Australia and the US) do not intend to ratify it (Yamin and Depledge, 2004). One interviewee from a non-Kyoto party expressed the view that 'the response to the US withdrawing from Kyoto has been managed in a way that has left the secretariat to be on one side of the fence... it's not quite the silent hand that was much more evident in the past when no one could pin down secretariat politics'. Certain public information documents produced by the secretariat have provoked private accusations of bias for assuming that the Kyoto Protocol is the way forward for the regime. The perception that the secretariat's support for the Kyoto Protocol constitutes partiality is problematic, given that the Protocol is a legal text of the regime and has been ratified by the vast majority of the Convention's parties. The dilemmas faced by the secretariat in this regard are likely to become more acute with the entry into force of the Kyoto Protocol.

The veil of legitimacy

An important means of managing the tensions inherent in the work of the secretariat is its symbiotic relationship with the presiding officers (or bureau), through which the secretariat carries out its activities and presents proposals *under the responsibility of these elected officers*. In doing so, the secretariat covers its own actions with a 'veil of legitimacy' of approval from the elected officers. This veil of legitimacy is absolutely fundamental to the secretariat's work. As discussed further below, the secretariat provides copious amounts of organizational energy to the negotiation process, making not only procedural proposals on how to manage the negotiations, but also substantive proposals on how to resolve differences between parties. The only way in which the secretariat is able to make those proposals and put them into practice is to do so through the relevant presiding officer who, assuming s/he is persuaded of the merits of the proposal, will put it forward in his/her name. This is accepted practice. As discussed in Chapter 4, presiding officers themselves do not always have the expertise, and rarely have the time, to engage in either detailed substantive drafting or reflection on organizational matters. The secretariat thus performs a valuable function in doing the thinking that provides the basis for the exercise of leadership by the presiding officers. The presiding officers, however, will always have the final say in whether a particular approach – substantive or procedural – is taken.

The importance of the veil of legitimacy is illustrated by the citation below. This reproduces a response by China (echoed by others) to a suggestion made at an informal group during the Kyoto Protocol negotiations that the secretariat should be asked to merge written proposals on a relatively uncontroversial issue.

> If you ask the secretariat to do ... only compilation ... they will do it very perfectly. But *if you ask the secretariat ... to merge something, there is the danger you will put the secretariat in a very delicate position* ... I wonder ... if we might request that you yourself [*the informal group Chair*] *take responsibility* for directing the secretariat. In this way, it would be a *party ... a member of this negotiating process, not the secretariat*, who would be responsible. We would accept, anticipate and ... believe that the secretariat would facilitate your work ... but by *your taking responsibility*, we might overcome this hurdle (AGBM 7, 1997a, emphases added).

In the end, the secretariat did merge the proposals and the informal group Chair made no changes to the secretariat's draft, but the fact that he exerted formal oversight over the secretariat's work was critical to the eventual legitimacy of the merged text.

The effective functioning of the veil of legitimacy is dependent on the relationship between the secretariat and the presiding officers. Where the secretariat has a good relationship with the presiding officers, this is of great benefit to the process. A prime example here is the relationship between Chair Estrada and the secretariat during the Kyoto Protocol negotiations. The secretariat, and particularly the Executive Secretary, enjoyed an excellent relationship with Estrada, built up over a long period since Estrada's involvement in the INC. The secretariat and Estrada remained in regular communication, especially by e-mail, between negotiating sessions, exchanging information and ideas. The appropriate relationship between presiding officer and secretariat, and especially the importance of the veil of legitimacy, was clear to both players, which was critical in allowing the secretariat to move more freely within its space for action. Estrada did not hesitate, for example, to publicly defend the secretariat at any unfounded suggestion that it had acted improperly. One such case occurred at AGBM 8 (part II) on the eve of COP 3, where some developing country parties objected to the latest negotiating text, on the grounds that the G-77 and China proposal for single year targets had been excluded in favour of multi-year targets. Unusually, the secretariat itself was criticized in plenary. Estrada, however, rose to the defence of the secretariat, insisting that the negotiating text was issued under his name. Estrada's intervention was very important here in maintaining the veil of legitimacy surrounding the secretariat. Although the parties knew perfectly well that the secretariat had done the actual work on the text, they had to accept that the final product was issued under the Chair's responsibility, and therefore that the secretariat could not be accused of any bias. It was accepted that the Chair, unlike the secretariat, did have a legitimate role to play in exercising leadership and putting forward proposals that he thought would eventually command consensus.

The relationship between secretariat and COP Presidents tends to be more complicated than with subsidiary body or informal group Chairs, as the COP President will often have a larger retinue of national advisers, who may give conflicting advice to that of the secretariat. This happened most noticeably, albeit to varying degrees, at COP 3 in Japan, COP 4 in Argentina and COP 8 in India. An interesting case concerns COP 6, under the Dutch Presidency, when cooperation between the secretariat and COP Presidency team improved considerably between the failed COP 6 and the successful COP 6 (part II).

Even outside the immediate negotiation process, the secretariat has been extremely cautious in the management of its public face. It has, for example, adopted a rather minimalist approach to public information, producing only a small number of guides and information sheets on the issue of climate change and the political process. This contrasts with the much greater public information campaigns of UNEP on behalf of the ozone secretariat,[3] which has prepared a considerable body of newsletters, booklets, leaflets, posters, calendars and other material on the ozone layer. The climate change secretariat, however, has been much more restrained, aware of the sensitivities surrounding the presentation and interpretation of the severity, causes of, and potential solutions to, climate change.[4] This approach is reflected in the New Delhi work programme on education, training and public awareness (decision 11/CP.8). This five-year programme of work is based on a 'country-driven approach' and focuses strongly on actions at the *national* level. Although the role of IGOs, NGOs and community-based organizations is highlighted, the secretariat is confined to a facilitative role, compiling and coordinating inputs from parties and the other organizations, and setting up an information-clearing house. The secretariat is certainly not expected, by this decision, to be producing its own expanded range of public information materials.

A great deal of effort has been expended by the secretariat to ensure the absolute objectivity of its information materials, within the framework of accepting climate change as a serious reality and the UNFCCC as the starting point for tackling it. Secretariat staff have been extremely reluctant to paraphrase or simplify complex negotiated text (e.g. the Convention, Protocol, COP decisions), even to make it more accessible to non-experts, for fear of inflaming sensitivities and being accused of bias. Interpretation and assumption have also proved dangerous. In an early version of one of its popular guides to the Convention and Kyoto Protocol, for example, the secretariat suggested, in a tabular chronology of the history of the process, that the Kyoto Protocol might come into force in 2002 (with two question marks to highlight the tentative nature of this). This reflected the aspirations of most governments that the Protocol should come into force in time for the World Summit on Sustainable Development (WSSD) (Johannesburg, August–September 2002) and the anniversary of the Convention's adoption, and linked up with the expectation that negotiations on the Buenos Aires Plan of Action would conclude by 2000. A government, however, raised an objection to this, stating that the entry into force of the Kyoto Protocol was a matter for the parties, not the secretariat, and the guide had to be modified.

Confidence and trust

Judicious use of the veil of legitimacy and caution in its public face means that the secretariat has long enjoyed a reservoir of trust among the parties in both its impartiality and effectiveness, which has translated into confidence in the negotiation process itself throughout its history. While there have been times when the secretariat has made mistakes, or its actions have been questioned (see examples given below), the parties have generally adopted a hands-off approach to the secretariat, trusting in its competence and integrity. This trust has built up since the secretariat's establishment due largely to the judgment, intelligence and personal affability of the first Executive Secretary. It is significant that Michael Zammit Cutajar came from a UN background, having risen up the ranks of the UNCTAD secretariat. He therefore 'brought with him excellent knowledge of the people in the UN system and the necessary grasp of complex UN procedures' (Kjellen, 1994, p152). His UNCTAD experience endowed Zammit Cutajar with a deep understanding of, and sympathy for, developing country concerns. His Maltese nationality was also an asset, with Malta being, at the time, both a member of the G-77 and a state aspiring to EU membership. Over his decade as Executive Secretary, Zammit Cutajar built up a good relationship based on respect and trust with all parties, even the obstructionists among them, knowing how to respond effectively to attacks on the secretariat or the process.

The tribute to Zammit Cutajar adopted as a resolution on his retirement at COP 7 illustrates the great respect in which he was held by the parties, referring to 'his achievement in building and leading an efficient and respected secretariat...his fairness and objectivity...commitment, professionalism and acumen' (resolution 2/CP.7).

To a large extent, the secretariat has retained this reservoir of trust following the handover to the new Executive Secretary, Joke Waller Hunter (the Netherlands). Waller Hunter's appointment by the UN Secretary-General was widely applauded as a wise move, and the change in leadership at the top of the secretariat has been relatively smooth. Waller Hunter appears to be succeeding in upholding the secretariat's good reputation. One interviewee commented on how she had 'managed to provide the secretariat with real credibility in difficult times...she deserves much credit for this'.

There are signs, however, that the long-standing confidence in the secretariat can no longer be taken for granted. At COP 9, for example, the G-77 and China called for a continuing review of the function and operations of the secretariat. The programme budget for 2004/2005 also includes a clause requesting the Executive Secretary to 'specify' how COP decisions relating to Article 4.8 – that is, the adverse effects of climate change and response measures on developing countries – are reflected in the work programme, along with a further request to conduct an 'internal review' of the secretariat's activities and report thereon to the COP (decision 16/CP.9). These clauses suggest the desire of parties, or at least the developing countries, to exercise more scrutiny over the secretariat's activities.

As part of the continuing review, initiated at SBI 20 in 2004, the G-77 and China raised questions over the geographical representation of secretariat staff, while calling for 'neutrality in preparing documents, and equity in the allocation of resources for issues concerning developed and developing countries' (ENB, 2004, p12). The Executive Secretary herself also faced criticism over statements made at public events that were interpreted as biased by certain parties, notably Saudi Arabia. The fact that these challenges are emerging now is certainly prompted, at least in part, by genuine concerns, notably over the low representation of developing country nationals among top secretariat staff. Undoubtedly, however, it also reflects opportunism by certain parties who see the secretariat as a vulnerable target in the pursuit of their political objectives. In this respect, the appointment of a new Executive Secretary from an industrialized country (however competent she may be) and the loss of the established relationship with Zammit Cutajar has provided parties wishing to do so with a window of opportunity to attack the permanent force of the regime.

Supplying organizational energy to the negotiations

The secretariat can, and does, supply organizational energy to the climate change negotiations in many different ways. These are discussed below.

Logistics

The most basic component of the organizational energy supplied by the secretariat is a logistical one, performed primarily by the conference services department, in collaboration with the UN Office at Geneva (UNOG). This includes, for example: ensuring the supply of sufficient appropriately sized meeting rooms; providing interpretation facilities into the six UN languages; processing documents, including translation and distribution; making venues available for informal and unofficial meetings; and identifying spaces for NGOs, other observers and indeed parties to hold side events and exhibits (see Yamin and Depledge, 2004).

The secretariat's logistical work is critical to the management of complexity, and the alleviation of procedural inequality among participants. It is the responsibility of the secretariat, for example, to ensure that information is freely accessible on the scheduling and location of meetings and on the availability of documentation, in order to maximize the opportunities open to all delegates to participate in the negotiations, as well as to minimize transaction costs. Delegates are kept informed primarily through a printed daily programme of meetings. Reflecting the growing complexity of the negotiations and the multiplicity of events taking place at any one time, since mid-1997 electronic noticeboards and (more recently) CCTV have displayed a continually updated full schedule of formal, informal and side event meetings. Since 1998, the daily programme of meetings has also provided a summary of the status of negotiations on all agenda items, indicating where an item is being worked on (e.g. contact group, informal consultation, plenary) and which presiding officer is responsible for the negotiations.

The importance of good logistics in a negotiation is often underestimated. However, in a complex, global multilateral process, if logistical issues 'are not properly addressed, the whole machinery gets out of order and no meaningful results can be achieved' (Kjellen, 1994, p152). At COP 6 (part II) and COP 7, for example, the conference services department was required to prepare rooms for high level meetings of the friends of the Chair to very detailed specifications at just a few hours' notice, including the provision of security, unique entry passes, special country flags and unusual seating arrangements. That they were able to do so in the required time frame was extremely important to maintaining the momentum of the negotiations. As a senior secretariat staff member noted in his interview, 'no meeting ever succeeded because the logistics were great. But if the logistics are bad, the negotiations can fail.'

Procedural management

Although the presiding officers retain authority over the negotiations as a whole, in practice, the main responsibility for the day-to-day procedural management of the negotiation process lies with the secretariat. Within the secretariat, this responsibility falls on the coordinators of the COP, SBSTA and SBI, and their professional staff, working with the conference services department on logistical matters. The secretariat supports the presiding officers in their exercise of process-oriented leadership, while taking the initiative to actively devise and propose ways of organizing the negotiations to promote agreement. The actions of the secretariat in this regard cannot be described as leadership, as its mandate forecloses, or at least heavily circumscribes, overt, independent action. The concept of organizational energy, however, captures well the way in which the secretariat injects strategic thought into the procedural management of the negotiations, but hiding behind the veil of legitimacy and shunning the explicit role of leader.

Prior to each negotiating session, the secretariat teams working for each body – COP, SBSTA and SBI, also the AGBM during the Kyoto Protocol negotiations and the JWG on compliance during the post-Kyoto process – make strategic suggestions to their presiding officers on how to organize the forthcoming session. They do so through informal e-mail and tele-communication exchanges, and briefing meetings held immediately before the negotiating session. Contacts are more formal and intensive in the case of the COP President, given the particular import and complexity of COP sessions and the typical lack of experience of Presidents in organizational matters. Suggestions usually cover the following issues:

- possible outcomes on each agenda item within the overall goal of furthering the regime
- the way each item could be dealt with – in a contact group, informal consultation, plenary discussion, or other arena
- ideas for delegates who might be invited to chair informal groups
- identification of potential procedural and substantive pitfalls and of how to deal with them, e.g. controversy over the provisional agenda
- scheduling of plenary and informal group sessions, and allocation of interpretation time
- arrangements for participation by NGOs
- correct protocol for dealing with ministers and other high level participants.

At the negotiating session itself, the secretariat supplies each presiding officer with speaking notes to help them chair meetings in a procedurally sound and effective manner. In addition, at least two professional officers typically sit alongside presiding officers on the podium during meetings, ready to provide them with any procedural and substantive support that they might need. This might include technical advice on the merits of different proposals, or how to respond to procedural concerns (e.g. points of order). Whether and how presiding officers draw on the support offered by the secretariat depends upon their confidence and experience in chairing meetings. Some presiding officers develop their own chairing style and just use the secretariat briefing notes as an aide memoire, while others read them out to the letter. Non-native English speakers, even if they are highly competent chairs, often find it useful to draw on the speaking notes of the secretariat in cases requiring a judicious choice of words (e.g. when seeking a mandate to prepare a negotiating text). Strong support by the secretariat has been particularly important for COP Presidents. The secretariat's ability to compensate for, and often paper over, the difficulties faced by many COP Presidents in the chairing of plenary and bureau meetings has been crucial to the smooth running of COP sessions.

Specific examples of organizational energy supplied by the secretariat in relation to the conduct of business and decision-making, negotiating arenas, negotiating texts, time management, and arrangements for participation of ministers and NGOs, are discussed in the relevant chapters of this book.

Being able to provide such organizational energy requires the secretariat to include among its staff individuals who possess experience in the organization of global negotiations in general, and experience of the climate change regime more specifically. In this regard, the continuity in the higher echelons of the secretariat, not just the Executive Secretary, but also the Coordinators of the SBSTA and COP (less so the SBI), has been important in building up the institutional expertise and memory of precedents that is crucial to institutional learning, and to the ability to think of creative options that can improve the management of the process.

Substantive input

Unlike the almost legendary role played by the UNEP secretariat in early negotiations under the ozone regime (see Benedick, 1991), the climate change secretariat has very rarely attempted to exercise open substantive leadership by brokering agreements among parties. This reflects the diplomatic, behind the scenes approach of the first Executive Secretary, which was itself a response to the desire of parties for a less interventionist secretariat than that of the early ozone regime.

However, the secretariat does have an extremely important role to play in putting forward ideas and draft language that might help reach a compromise. These are almost always put forward through discreet advice to the relevant presiding officer, thus maintaining the veil of legitimacy. For example, at COP 3, the secretariat gave advice to Chair Estrada on the numbers he should propose in a first draft list of emission targets, based on the secretariat's understanding of party positions. As one interviewee, who had served in a chairing role, characterized it:

> Their role is not to get involved in the negotiation. But I do see the secretariat as someone who...will always be there to assist the Chair...by providing him some suggestions...Just because you are not involved in it, and you are sitting at a distance, you can also identify areas of convergence...and you can tip off the Chair – look, that's where you can move.

Drafting text

This substantive energy provided by the secretariat is particularly important in promoting consensus through the medium of drafting text for presiding officers (including Chairs of informal groups) to consider presenting as their own. Almost every decision or conclusion that is negotiated in the climate change regime is based on an original draft, or compilation, suggested by the secretariat. Draft decisions and conclusions are often circulated and discussed internally within the secretariat prior to a negotiating session, and presented to the relevant presiding officer, or even shared with trusted parties. Professional staff working for each body mostly include individuals with the drafting skills and technical expertise needed to manipulate the subtleties of the English language to find the right diplomatic wording. The secretariat is also able to draw on its institutional memory of precedents used in the past to help craft acceptable language.

The work of the secretariat in drafting text is an important dynamic, helping to move the negotiations forward. For one interviewee, commenting on the Kyoto Protocol negotiations, 'the secretariat played an extremely substantive...extremely forceful role in the development of the text, getting...significant text out that... reflected all the views on the table'. The drafting work of the secretariat also reflects an efficient division of labour, freeing the presiding officers (many of whom may be neither native English speakers nor good drafters) to apply their political skills to the substance of negotiations. As one interviewee explained, again with reference to the Kyoto Protocol negotiations, the Chair 'clearly wasn't going to be able to come up with drafts on absolutely everything just off the top of his head [he], was able to rely on the production of written text [by the secretariat] which he just wouldn't have been able to handle all by himself'.

Technical advice

Another area of substantive support by the secretariat is in the provision of technical expertise to help parties take the most effective decisions. The secretariat does so in three main ways:

- writing official technical and background documents, mostly on request from the COP or subsidiary bodies
- providing technical advice to the presiding officers, either through written briefing notes on specific issues or orally during the negotiation process
- providing technical advice directly to the parties, usually informally in the corridors. It is relatively rare for the secretariat to be formally invited to give technical advice during an official meeting, even in an informal group.

The secretariat has, on occasion, been extremely important in supplying technical analysis on new issues in the negotiations. Such scoping work has then served as a basis for parties to begin negotiations or to develop their own positions. Input from the secretariat in this way can be particularly significant in helping to build up a common base of knowledge and understanding among parties, in a context of differing capacities among countries. In the follow-up to COP 3, for example, the secretariat – formally on request from the subsidiary body Chairs – prepared papers analysing the Protocol's three flexibility mechanisms and LULUCF provisions, and identifying issues for further work. These two documents were important contributions to framing discussions at the first negotiating sessions after COP 3, when parties were rather at a loss on how to proceed. Interestingly, the paper on the LULUCF sector was better received than that on the mechanisms. This can be largely attributed to the fact that the LULUCF paper was more scientifically factual and technical in nature, and could draw on existing literature. The flexibility mechanisms paper, however, dealt with an entirely new set of concepts, and therefore the analysis of the secretariat was inevitably more open to challenge and accusations of bias.

The period since the adoption of the Kyoto Protocol has seen a rise in the technical aspects of the secretariat's work. The need for more complex emission inventories and accounting methodologies, the inclusion of the LULUCF sector, and unprecedented technical issues arising from the flexibility mechanisms, have all created greater demand for technical expertise on the part of the secretariat. In this respect, the Kyoto Protocol negotiations contrast sharply with the post-Kyoto phase. The Kyoto Protocol negotiations were managed within the secretariat by a small team of political experts with little technical expertise, so that, while the presiding officers received important strategic advice, technical input was limited. In the post-Kyoto negotiations, however, each specific issue was managed by a team with expertise in that field, so that the level of technical input was much greater. Lack of analysis of technical issues during the Kyoto Protocol negotiations meant that the secretariat failed to bring certain important substantive matters to the attention of the Chair in a timely manner, in particular the treatment of the LULUCF sector, and also the implications of the various proposed emission targets. One interviewee commented, 'the sinks issue is one where I feel the secretariat could have had a much stronger role in pinpointing to Estrada – who was, after all, a policy, non-technical person – that there was a need to address this issue in more detail'. Interestingly, the former Executive Secretary himself acknowledged that the Kyoto Protocol negotiations had lacked technical support: 'What could we have done differently? Data issues, we could have realized the importance of this before the last minute, and work on sinks. I didn't understand the need for data work... If I had to do the Kyoto Protocol now, I would allow more capacity for technical work.'

Facilitating informal, in-depth discussion

A related means by which the secretariat has put substantive energy into the climate change regime has been the holding of side events (see Chapters 10 and 14) to facilitate more informal, in-depth discussion than is possible in the official negotiating arenas. Most of the side events organized by the secretariat are technical in nature (e.g. briefings on the reporting guidelines, or on the secretariat's work on technology transfer), or provide a forum for parties themselves to discuss issues and exchange experiences. Side events where non-Annex I parties present their newly submitted national communications, for example, have now become a tradition in the climate change calendar.

In some cases, the secretariat has arranged more exploratory side events to help advance discussion on new or particularly complex issues in the negotiations. During COP 4, for example, the secretariat ran a special event on issues and options for the CDM, providing a valuable informal forum for substantive exploratory discussion on what was then a very new concept. A similar special event was held at COP 5 on options for the treatment of liability under emissions trading. In another example, at SBSTA 12, the secretariat organized a special event on education, training and public awareness, in order to facilitate an initial discussion prior to the start of negotiations on the topic.

A pioneering recent initiative by the secretariat has been the organization of high-level events, usually in collaboration with businesses and other NGOs, to encourage a broader discussion of climate change related issues beyond the immediate agenda. At COP 9, for example, the secretariat coordinated high level side events on the transport sector, enabling environments for technology transfer, and opportunities offered by the CDM to provide clean and affordable electrical power to the developing world. These events brought together senior representatives from business, environmental NGOs (ENGOs) and governments, providing an opportunity for informal interaction 'outside the box'. Another emerging tradition is the organization by the secretariat of very informal discussions at the secretariat's information desk, the climate change 'Kiosk'. These involve soapbox-type presentations made by a range of NGO speakers on broad climate change-related topics. Topics covered at COP 9, for example, included mitigation/adaptation, global climate science and transport.

These side events coordinated by the secretariat have proved useful in encouraging wide-ranging discussion and forward thinking beyond the constraints of the formal negotiation process. They have been particularly helpful in encouraging partnerships between different stakeholders – notably ENGOs and business and industry NGOs (BINGOs) – while endorsement by the UNFCCC Executive Secretary has helped ensure high-level participation, and therefore a high profile. While outside the formal negotiating structure, such events are indicative of an active and engaged secretariat, and help to inject important substantive energy into the regime.

The dangers of going too far

There are several instances in the history of the secretariat where its approach to providing substantive input has gone too far, or has otherwise been perceived as inappropriate by the parties. This limited number of cases is instructive in illustrating the importance of the veil of legitimacy, along with the fragile balance that the secretariat must constantly maintain between providing much-needed advice and technical input on the one hand, and maintaining absolute objectivity on the other.

One case concerns the debate on sinks during the Kyoto Protocol negotiations. These negotiations only began towards the end of the process, partly because the secretariat team had failed to appreciate the import of the issue until very late on. The negotiations on sinks were supported by a senior staff member with some technical knowledge of the issue, who adopted a more interventionist stance, consulting actively with parties behind the scenes. Some parties were unhappy with this approach, although concern was never publicly aired. One interviewee commented:

> [the secretariat] played a much more activist role ... in the area of sinks. ... I'm not sure that worked ... it's very hard for the process to separate technical expertise from political views when they come from an ostensibly neutral source. And I think those often overlapped ... you always have people in senior positions, and they have views, and I'm not persuaded that in the sinks discussion that was helpful.

Another case occurred at COP 6 (part II) where, following triumphal 'approval' of the Bonn Agreements by the COP plenary, the secretariat was given a mandate to edit the text and adjust 'a number of legal and technical points', prior to its translation and publication as an official COP document (see COP 6 (part II) report, paragraph 39). The secretariat, however, went too far in seeking to tidy up the text, introducing changes which, while they improved the sense and consistency of the language, were deemed to have political implications. This excessive zeal was due in large part to the personal sense of commitment of secretariat staff to the post-Kyoto negotiations, on which they had worked long and hard for years, and their desire to produce a rigorous, precise and elegant text. Although the document was embargoed, it was inadvertently released, causing consternation among delegations who were nervous of the potential political repercussions of the ostensibly 'technical' changes. The blunder was patched up by withdrawing the document, and reissuing another version reproducing exactly the text approved by the COP, including obvious technical and editorial mistakes. There were also more long-standing repercussions, as the COP Bureau was then reluctant to entrust the secretariat with editorial tidying up of the Marrakesh Accords at COP 7.

A further example concerns a document issued at SBI 19 (held in conjunction with COP 9) in 2003, providing 'an initial assessment of steps taken by non-Annex I parties to reduce emissions and enhance removals of greenhouse gases' (see FCCC/SBI/2003/INF.14). This document was prepared as a 'note by the secretariat'. That is, responsibility for its contents lay with the secretariat, as opposed to a 'note by the Chair', where the responsibility would rest with the Chair, even if the secretariat had actually drafted the text. Although the secretariat had been mandated to work on this topic, the document that was presented to the SBI was vehemently criticized by the G-77 and China, and had to be withdrawn before developing countries would agree to adopt the SBI agenda at the opening of the session (some industrialized countries, however, notably the US, applauded the document). Senior secretariat staff were blamed for not having sufficiently appreciated the sensitivities involved in drafting a document on such a controversial topic, and the acrimonious debate among parties surrounding the document coloured the whole debate on non-Annex I national communications at that session. It also provided an opportunity for certain laggard developing countries to demand that the secretariat's work be more closely monitored by the parties, suggesting the beginning of an overall erosion of confidence in the secretariat (see above). This illustrates how careful the secretariat has to be in its presentation of issues, and indeed how difficult that job can be.

Summary and concluding remarks

This chapter has outlined various ways in which the secretariat has supplied organizational energy to the negotiation process, with other examples appearing throughout the chapters of this book. A key point to note is that the UNFCCC secretariat has, to date, been held in high regard by the parties, for its competence, efficiency and objectivity. This has exerted an immensely positive influence on the overall negotiation process. Although some challenges are now being raised to the secretariat, these are largely opportunistic and, given the large reservoir of trust it enjoys, are unlikely to prevent it from continuing to input organizational energy to the process.

The ability of the secretariat to contribute meaningfully to the process depends, perhaps paradoxically, on its invisibility. Throughout the history of the climate change negotiations, the secretariat has not been interventionist; however, it has been highly active, working through the presiding officers behind the scenes where its activity is invisible to most delegates. One interviewee captured this point eloquently:

> The secretariat is held in very high esteem by the parties, and... justifiably so. And part of that is because of their consistent exercise of self-restraint... you don't ever get the impression that the secretariat... is manipulating the process to produce a particular outcome... they probably have more influence through that almost subliminal assistance.

The experience of the climate change secretariat points to several main attributes that can contribute to a secretariat's effectiveness, beyond an adequate financial base and political support. These are rather obvious, but are worth restating. Continuity of (competent) senior staff is vital, especially given the very complex politics of the climate change regime. Continuity enables senior staff to develop personal relationships with delegates, and also to build up the base of historical experience and knowledge of precedents that is needed to devise workable, politically acceptable solutions to organizational challenges. Also vitally important is achieving a good mix among secretariat staff between technical expertise and political acumen. It is not always possible to find these two qualities in a single individual, but it is necessary for the secretariat, as a whole, to possess a balance between them. By and large, the climate change secretariat has succeeded in doing so.

The secretariat's role has developed in parallel with the climate change process itself. As one secretariat interviewee put it: 'The secretariat's role has changed, continues to change, and we still don't know how this will play out. It is evolving as the process has evolved.' In this respect, a recent trend has been the greater specialization of particular secretariat staff and departments to a focus on specific issues (e.g. LDCs, technology transfer) as the climate change process itself has become more fragmented. The focus of the secretariat's work has also become more technical, in keeping with the implementation phase of the post-Marrakesh process.

The entry into force of the Kyoto Protocol, and the new implementation phase of the climate change regime, means the secretariat finds itself at a propitious point to consider its future. Whether the secretariat could, or should, assume a more proactive role in the regime is a question that is being asked both within the secretariat itself, and among some parties. The most promising area would be in terms of technical analysis, identifying issues and proposing, perhaps even recommending, solutions in a more forthright manner than before. Public information is another, but more difficult, area, given the controversy that still surrounds the issue of climate change, and the existence of powerful groups who remain sceptical. One secretariat interviewee raised the question as follows: 'Previously the secretariat was so careful – staff were reluctant to paraphrase anything at all. But now they are starting to ask themselves how far they can speak. Can we issue press releases? Respond to questions on climate change? Can the secretariat be an authority on the topic?'

Perhaps the most important question is the extent to which the secretariat could assume a leadership role in the event of deadlock over the future of the climate change process and the absence of effective leadership on the part of any party. In such a situation, of major parties unable to agree on a way forwards over future commitments, the distinction between process-oriented and substantive leadership would be critical. The secretariat would not be able to take the lead in promoting a particular way forwards (e.g. negotiation of stronger emissions targets for all parties, or just Annex I parties). However, it might be able to help unblock a stalemate through the presentation, perhaps under a mandate from the Bureau, of different options and directions in which the process could move (including the option of doing nothing). The secretariat could thus exert true process-oriented leadership by at least identifying a stalemate and posing the issue of what to do next, and then help focus the minds of parties on the options and their implications. This, in itself, would be a very valuable function. The secretariat would be unable, however, to exert genuine substantive leadership by attempting to steer parties in a particular direction, without forfeiting its greatest asset – trust and confidence in its objectivity.

Rules for the Conduct of Business

> To the outsider, procedures may seem peripheral to matters of substance, but since in the UN procedures can determine substantive outcomes, they can and do arouse great passions (Renninger, 1989, p235).

Introduction

This chapter explores the rules for the conduct of business that govern the actions of all the players in the climate change negotiations – both parties and organizers (presiding officers, bureau and secretariat) – woven around the formal Convention text and draft rules of procedure, along with the informal practices and institutional setting of the regime. The formal rules are largely unoriginal in the climate change regime, mirroring those at work in the wider UN system, with few concessions made to the uniqueness of climate change. What is interesting, however, is the way in which the players use and interpret the rules, how they improvise around them in innovative ways, and how they harness them to promote their goals in the negotiations. In this chapter, we begin by outlining the key rules and informal practices that shape the conduct of business, before examining how these have been applied in practice in the climate change negotiations.

Key rules and informal practices

The key rules and informal practices relating to the conduct of business in the climate change regime include the following:

- The foundation for the conduct of business is the *one-party-one-vote* provision, set out in both the Convention and the rules (Article 18 and Rule 41), which is standard in most UN forums. Although there are few instances in which votes can be taken, by extension, this provision translates to 'one-party-one-voice', in other words, 'equal say' for all parties.

- Debates cannot proceed 'unless at least one third of the parties... are present'. This required *quorum* is increased to two-thirds for decision-making (Rule 31).
- Rules governing the *right to speak* establish that delegates may only speak when called upon by the presiding officer, who should call speakers in the order they ask for the floor, based on a list of speakers kept by the secretariat. Although the presiding officer may call speakers to order if their remarks are not relevant, s/he must seek the agreement of the parties to limit the overall time, and the number of times, a delegate may speak (Rule 32).
- Statements made in an *official language* of the UN (Arabic, Chinese, English, French, Russian, Spanish) must be interpreted into the other five (Rule 55). Official documents must likewise be drawn up in one language, and translated into the others (Rule 56).
- *Documents should be circulated in advance.* Supporting documents for a negotiating session must be distributed to parties at least six weeks before the opening of that session (Rule 11). During a session, no proposal is to be discussed unless copies have been circulated to delegations the day before, although the presiding officer may exercise discretion and waive this rule (Rule 36). Special rules apply to the text of any proposed amendment, annex or protocol which, according to both the Convention and the draft rules of procedure, must be communicated to parties at least six months before its adoption (the 'six-month rule') (Articles 15, 16, 17; Rule 37).
- *Standard meeting hours* are 10:00–13:00 and 15:00–18:00, while *no more than two official meetings*, including formal plenary meetings and informal groups, should meet at any one time. These are well-established informal practices, derived from the institutional setting of the wider UN system.

Functions of rules and informal practices

These rules[1] for the conduct of business are aimed at ordering debates in a predictable and standardized manner, thereby serving as an indispensable anchor to manage the complexity of global negotiations. Without accepted rules to govern the exchanges among parties, global negotiations would descend into chaos. The rules thus help to reduce transaction costs in very practical ways, for example, by ensuring familiarity among parties as to how a formal subsidiary body or COP meeting will proceed, its scheduled start and end times, procedures for intervening in debates, expectations regarding interpretation, and so on.

Aside from the standardization of the conduct of business, the rules lean more towards the promotion of procedural equity and transparency in the negotiations, than the maximization of efficiency. The most obvious example is the equal say rule, which enshrines formal procedural equity by giving all parties an equal right to contribute to debates and decision-making.[2] The presiding officer is expected to call on speakers seeking the floor through the objective procedure of first-come-first-served and, once a speaker has begun to speak, s/he cannot be interrupted by other parties (who may, for example, be more eloquent or better informed), as these must seek the floor before intervening. Although the formal rule whereby the presiding officer has to seek the consent of parties to limit the speaking time of a delegate is very rarely invoked, the associated established informal practice in the UN system whereby a party is permitted to speak without restraint is treated with respect. This is particularly important for developing countries, who are thereby provided with a valuable forum to air positions that they feel are often ignored by their more powerful counterparts in the industrialized world. As Hyder comments:

> The UN system permits all sides to express their opinions from a position of sovereign equality and, therefore, to maintain self-respect. Countries acknowledged to have dominant economic, political and military power are forced to take into account the contrasting views of many other countries, however weak those other countries may be (Hyder, 1994, p203).

Given its widespread application in the UN system, the equal say rule is largely taken for granted in the climate change regime. A Chinese delegate, however, illustrated the underlying importance of this provision when he affirmed, during the final stages of the Kyoto Protocol negotiations: 'We [the G-77 and China] will not be pressured into accepting this proposal. This is not Bretton Woods. This is the United Nations' (CoW, 1997d). This statement made the point that, thanks to the equal say rule of the climate change regime, developing countries were negotiating on a procedurally equal footing with the industrialized countries, unlike the Bretton Woods institutions of the World Bank and IMF.

Other rules for the conduct of business seek to give practical effect to formal procedural equity by promoting a level playing field, that is, trying to compensate for differing resource levels and negotiating capacity among parties. Interpretation and translation requirements, for example, help to counter the massive disadvantage faced by non-Anglophone delegates and is of particular use to developing countries and EITs who rarely have the resources to provide their own language services. Advance circulation of documents, in turn, seeks to give all delegations, especially small ones with limited analytical resources, adequate time to consider proposals before being required to discuss or decide upon them. The established informal practice of no more than two meetings is similarly intended to maximize opportunities for small delegations to participate in negotiations. A key practical mechanism for seeking to uphold formal procedural equity is the funding of participation at negotiating sessions by eligible developing countries and EITs (see Box 7.1 below).

The rules also have an important transparency dimension, with the main rules in this regard being the quorum requirements and the public nature of meetings. In addition, minimizing parallel negotiations through the no-more-than-two-meetings practice, broadening the accessibility of proceedings and

Box 7.1 Funding for participation

In recognition of the more limited resources of most developing countries (and EITs), many environmental regimes provide funding to support their physical attendance at negotiating sessions. In the climate change regime, a Trust Fund finances the participation of one delegate from each eligible developing country and EIT per negotiating session. If funds are available, support is provided for a second delegate from Small Island Developing States (SIDS) and LDCs to attend COP sessions. Eligibility for funding is determined according to GDP per capita. The current threshold in the climate change regime is that GDP per capita must not have exceeded US$ 6,500 in 2000 according to the Data Management Service of UNCTAD, rising to US$ 10,000 for SIDS and parties providing Bureau members (see Yamin and Depledge, 2004).

Funds are drawn from voluntary contributions, mostly from industrialized countries. The voluntary nature of contributions introduces uncertainty as to the funding that will be available to cover participation. For the most part, there have been sufficient funds (although only just) to finance the participation of at least one eligible delegate at each session. The small number of exceptions, however (including at AGBM 1 during the Kyoto Protocol negotiations and at SBSTA/SBI 18 in the post-Marrakesh era) prompted strong expressions of concern by the G-77 (see also Yamin and Depledge, 2004).

Providing such financial support for participation is a widely accepted part of the organization of the negotiation process, with its practical benefits appearing straightforward and indisputable. Funding for participation also has a symbolic dimension that is important to enhancing the legitimacy of negotiations. Most interviewees approved of the provision of such support One respondent suggested that, given the multi-issue nature of climate change, 'support for non-Annex I parties should enable participation by at least three participants from different disciplines, such as economics, science and law'. Overall, as one interviewee explained, 'there's still a huge inequity in the participation of developing countries ... because developed countries can just have resources to bring in as many people as they can'. The climate change regime can clearly only go so far to redressing the broader inequities of the international system.

Some interviewees, however, mostly from Annex I parties, expressed misgivings. One interviewee commented that providing developing countries with funding conveyed the impression that the negotiations were not of their concern and they had to be paid to attend. Several interviewees also noted that funding for participation was sometimes associated with rent-seeking activity. One interviewee noted that 'the amount of DSA [daily subsistence allowance] can amount to six months salary. There is fierce competition to get to meetings.'

The issue underlying the above comments is that funding physical attendance does not necessarily equate with active participation. Increasing the number of individuals on a delegation, or enabling a country to be represented that would not otherwise be, is an important first step. However, it does not guarantee that those delegations will then have adequate knowledge of the issues, or the right skills, or a sufficiently clear mandate, to participate effectively in the negotiations.

documentation through interpretation and translation, and allowing time for the consideration of proposals before decisions are taken, all also increase opportunities for scrutiny of the negotiation process by both parties and non-state organizations.

In promoting procedural equity and transparency, however, the procedural rules also tend to raise the transaction costs of negotiating. Interpretation requirements, the standard meeting schedule and the no-more-than-two-meetings practice, for example, all limit the time available for negotiation and opportunities for convening a meeting at short notice. Rules concerning translation and advance distribution of documents are also potential curbs to spontaneity. Moreover, formal procedures for seeking the floor, coupled with the discipline imposed by simultaneous interpretation, place obstacles to the uninhibited exchange of views, while openness to the public can encourage posturing and inflexibility (see also Chapters 9 and 14).

Procedure as strategy

The rules for the conduct of business of the climate change regime are frequently harnessed by parties as strategic means to pursue political objectives. This is not unusual. As Kaufmann notes, '"procedure" and "substance" … are supposed to be two separate things. In practice, however, procedural devices are used to obtain a substantial result and procedural debates often turn out to be debates on substance' (Kaufmann, 1989, p40–41).

It is not uncommon in the climate change negotiations for some parties, when faced with a proposal to which they object, to invoke procedural arguments for why it cannot be discussed (e.g. non-availability of a text in all languages, insufficient time to consider it). It is similarly not uncommon for some parties to seek to delay the negotiation process more generally by insisting on procedural adherence (e.g. on interpretation or translation, or strict application of the no-more-than-two-meetings rule) (see Yamin and Depledge, 2004). One interviewee claimed: 'If the Saudi delegate puts his request for translation in perfect English it's very clear that it's a trick and nothing else'. The line between legitimate and illegitimate procedural challenge is, of course, a fine one. Most procedural concerns are motivated by genuine concern over loss of procedural equity and transparency. Nevertheless, it is also undoubtedly the case that obstructionist parties sometimes insist on the rigid interpretation of procedural rules as a means of delaying the negotiation process, and it is no coincidence that certain OPEC countries tend to be the greatest sticklers for procedural adherence.

An interesting example arose during the Kyoto Protocol negotiations at AGBM 3 in early 1996, when Saudi Arabia put forward an interpretation of the six-month rule whereby there could be no further negotiation on the negotiating text of the protocol between its circulation and its adoption. In effect, this would have required the negotiations to be completed by 1 June 1997, an inconceivable prospect. This challenge was eventually headed off by a legal opinion from the UN Office of Legal Affairs, which confirmed that negotiations on the protocol could continue up to COP 3 (FCCC, 1996; see also Chapter 5). A more commonly used example of procedural opportunism occurred at SBSTA/SBI 13 in 2000, when Saudi Arabia, whose negotiators mostly possess excellent English language skills, exerted sufficient influence in the G-77 to demand, on behalf of the group, translation of all documents, an obvious physical impossibility. According to Grubb and Yamin, 'established UN procedures are slow and painful, and certainly in the history of the climate change negotiations they have been deliberately used by laggard countries to prevent effective international policy-making on climate issues' (Grubb and Yamin, 2001, p270). An interviewee similarly complained that, 'We are treated as fools here by delaying tactics'.

Putting the rules into practice

Given the higher transaction costs that they impose, the structure provided by the rules for the conduct of business is generally a rather loose one. The rules are used to structure and order the negotiation process, as well as to enhance procedural equity and transparency, yet at the same time they are applied pragmatically, regularly bypassed and, at times, relaxed, when the transaction costs that they impose threaten to impede the negotiations. It is extremely important, however, for the organizers of the negotiation process – the presiding officers, bureau and secretariat – to make sure that a balance is secured. Enough of the key rules for the conduct of business must be upheld enough of the time to maintain an acceptable level of procedural equity and transparency, if the legitimacy of the negotiations is not to suffer.

Pragmatic application

A good example of the pragmatic application of the rules for the conduct of business is that of the quorum. The presiding officers and secretariat typically verify that representatives of the main negotiating coalitions are present in the plenary room before starting a meeting, but rarely actually check the number of countries. Such a pragmatic quorum is in fact more important than the formal numerical threshold, which could be reached simply through the presence of the G-77, without any Annex I party. Transparency is thus maintained by respecting the spirit of the rules, while avoiding an unnecessary rise in transaction costs by adhering to their letter.

Another example is the rule requiring presiding officers to call on speakers 'in the order in which they signify their desire to speak', with the secretariat maintaining 'a list of speakers' (Rule 32.1). In practice, seating arrangements make it difficult for parties to keep track of who is asking to speak, so that the presiding officer can usually use his/her discretion to decide on the exact sequence in which s/he calls upon parties (Yamin and Depledge, 2004). This can be a very important tool in the negotiations. One presiding officer, for example, admitted in his interview that he took strategic decisions on the order in which to call upon speakers. Depending on the 'mood of the room', the statements of obstructionist parties, for example, might be taken later on, after support from others had been made clear, or alternatively they might be taken first, to be drowned out by subsequent supportive statements. Other presiding officers have also used this tactic. Chapter 8, for example, refers to a debate on LULUCF where only a handful of parties were opposed to convening an inter-sessional workshop on this issue. Chair Dovland first took a statement from the main opponent, Saudi Arabia, and then, even though Saudi Arabia soon asked for the floor a second time, instead gave the floor to the majority of parties expressing support for the workshop so that, when he finally did call on Saudi Arabia a second time, that party had little choice but to acquiesce.

Bypassing

The most common example of the bypassing of rules for the conduct of business is the convening of *informal arenas* where the rules (e.g. relating to interpretation, or meeting times) are either not applied, or applied only in part. It is indeed commonplace in the multilateral arena to refer negotiations to informal groups, which, as Iklé puts it, 'sometimes help to avoid the more irrational and inefficient aspects of the verbal exchange at the conference table' (Iklé, 1964, p118). The use of informal arenas has intensified over time (although there is no clear trend in the actual number of arenas convened – see Chapter 9). This is due partly to the greater number of items on the subsidiary body and COP agendas, but also to the almost automatic assumption that an issue, if it is to be considered seriously, must be addressed in an informal group or informal consultation. This has led to a situation whereby plenary discussion of agenda items has become increasingly perfunctory, as parties – and indeed presiding officers and the secretariat – take it for granted that the real debate will unfold informally. According to one interviewee, 'the use of both subsidiary body and COP plenaries are much more ritualistic, very little action happens there at all'. The result is that the vast majority of actual negotiations take place in informal arenas where there is no obligation to adhere to the full rules for the conduct of business. While more efficient, the multiplication of informal arenas does make it more difficult for small delegations to participate effectively in the negotiations. Issues relating to informal arenas are discussed in more detail in Chapter 9.

Another common case of bypassing the rules is through the issuance of *informal documents*, which do not need to comply with translation and advance circulation requirements (see also Chapter 11). These include information-only documents (INFs), conference room papers (CRPs) and miscellaneous documents (MISCs), which, as informal documentation, may be issued in English only, and at short notice (see Box 11.1 on document types). The extreme manifestation of the bypassing of documentation rules is the use of informal *and unofficial* 'non-papers', that is, text simply reproduced on blank paper, without an official UNFCCC symbol or logo. The best example of the bypassing of documentation rules concerns draft decisions and conclusions. When these are presented for adoption to a plenary meeting, they should be issued in a formal document in all UN languages. However, because decisions are usually agreed at the last minute, there is often insufficient time to translate the text. Therefore, to bypass the translation requirement, the text is issued as a CRP, in English only. The number of decisions and conclusions adopted as CRPs has risen dramatically since Kyoto, reflecting the greater amount of business and pressure on translation services. Parties have been remarkably acquiescent to this trend, generally recognizing and accepting the logistical impossibility of translating text agreed in the very last hours of a negotiating session.

Relaxation

As well as being bypassed through informal groups and informal documents, the rules are sometimes explicitly relaxed during formal meetings in the face of practical constraints. On occasion, for example, formal plenary meetings have extended beyond their scheduled meeting time or have been held without interpretation. It should be noted that available interpretation time is limited during negotiation sessions, due to both budgetary constraints and strict regulations governing the working hours of UN interpreters. Interpreters have sometimes continued to work beyond their allotted hours, but there is no compulsion on them to do so, and the extent to which they are physically able to carry on is limited. Similarly, the no-more-than-two-meetings practice has occasionally also been relaxed during intensive bargaining and deal-making in the finales of major negotiation rounds. Relaxation of the rules is, however, often contested, notably by the G-77 and China given that developing countries are most in need of their procedural equity and transparency safeguards.

The willingness of parties to agree to the relaxation of the rules varies, and is largely a function of expediency and political will. This can be illustrated by different examples from the Kyoto Protocol negotiations. At AGBM 2 in 1995, for example, Chair Estrada was forced to close a relatively insignificant meeting although he had not completed the list of speakers as parties refused to carry on for 15 minutes without interpretation (AGBM 2, 1995); this may have been because parties did not want to set a precedent so early in the process. As negotiations advanced and the need for efficiency increased, the G-77 and China showed considerable flexibility in agreeing to bypass and relax the rules. The Chair of the G-77, for example, acceded to Estrada's proposal during COP 3 to 'have a six hour meeting this evening without interpretation', stating that 'we, in the G-77 and China, are willing to stay on in order to complete the work' (CoW, 1997b). A related situation unfolded with regard to the no-more-than-two-meetings practice. The G-77 and China kept a close eye on the organization of the negotiations at AGBM 6 and 7 in 1997 when informal groups were first convened to ensure that only two meetings were held in parallel. During an AGBM 7 plenary meeting, for example, Iran noticed that a proposed scheduling change would mean that three meetings would take place in parallel, and immediately intervened to object (AGBM 7, 1997b). However, at COP 3, the G-77 and China implicitly acquiesced to the convening of multiple parallel informal groups by not raising objections to the schedule.

The most striking example of the relaxation of the rules concerns the adoption of the Kyoto Protocol itself, which took place in procedurally irregular circumstances. Due to the overrunning of COP 3 by almost a day, the final hours of negotiations in the Committee of the Whole (CoW, see Chapter 9) and the entire final COP plenary meeting, including the formal adoption of the Kyoto Protocol and other decisions, took place in English only. Moreover, there was probably no quorum to take a decision, as many developing country delegates had also left to catch UN-funded flights home. One interviewee recalled, 'I remember being alone in that room, I don't know how many delegations were there, but there couldn't have been more than a dozen.' However, these irregularities generated only one objection, from the Russian delegate over the lack of interpretation, who nevertheless did not insist on the suspension of the negotiations (CoW, 1997f). It was probably critical, however, that the final text of the Kyoto Protocol was available in all six languages for adoption, in accordance with the procedural rules. Whether delegates would have conceded to adopting the written text of the Protocol in English only, in addition to all the other procedural irregularities, is impossible to tell.

The interplay between procedure and political will

The Protocol negotiations thus show how *political will can override procedure*. When faced with practical imperatives, particularly time pressure, parties did consent – developing countries more reluctantly than others – to relaxing procedures even within formal meetings. When it came to the crunch, parties were prepared to sacrifice some procedural equity and transparency for the sake of efficiency and, ultimately, reaching agreement.

Since Kyoto, however, the willingness of parties to accept the pragmatic application, bypassing or relaxation of rules for the conduct of business appears to have waned. Developing countries in particular have become more assertive in insisting on the no-more-than-two-meetings practice and, with the Russian Federation, on calling for the suspension of formal meetings when interpretation is no longer available. Interpretation was even requested, and in the end provided, for some meetings of the larger contact groups, notably on the flexibility mechanisms, during the post-Kyoto negotiations. At SBSTA/SBI 13 in 2000, the last session in the post-Kyoto negotiations before the scheduled finale of COP 6, the Nigerian G-77 Chair repeatedly raised concerns in plenary at the lack of interpretation, paucity of translated documents, and late circulation of documents. In his opening statement, he also called (unsuccessfully) for a one hour break between each meeting and an end to the working day at 9pm. The COP 6 finale, in turn, contrasted with COP 3 in terms of the procedural flexibility shown by parties. The Chair of the G-77 repeatedly objected to proposals by the COP President for convening late night, parallel, or English-only informal meetings, greatly reducing the efficiency of the negotiations. The Russian Federation's strong insistence on provision of interpretation (see Chapter 13) also contrasted with that delegation's (albeit reluctant) acquiescence of English-only negotiations in the last hours of COP 3.

The situation changed again at COP 6 (part II), at least as far as the G-77 were concerned, with the Iranian G-77 Chair able to harness more flexibility on the part of his Group and voicing fewer procedural objections to English-only, late night, or parallel negotiations. This was a case, once again, where political will – in this case determination to avoid another failure like that of COP 6 and to uphold multilateralism – was able to override procedural concerns. Echoing the sentiment implicit at COP 3, however, the organizers still considered that it would be unacceptable formally to adopt the Bonn Agreements – finalized in marathon, overnight negotiations – in English-only as soon as they were agreed. Instead, using a procedural sleight of hand, the English version was immediately 'approved' by the COP plenary, 'on the understanding that *formal adoption* would follow at the next plenary meeting, once the text had been issued *in an official conference document*', that is, in all UN languages (COP 6 (part II) report, paragraph 40, emphasis added). The decision was duly adopted two days later in all languages.[3] In another illustration of the interplay between procedural adherence and political will, however, in the wake of approval of the Bonn Agreements, the Russian Federation insisted on closing a crucial COP plenary meeting when interpretation time ran out before important decisions on the organization of work could be agreed, despite this effectively putting a stop on negotiations for 24 hours.

It is difficult to dismiss the conclusion that foot-dragging played at least some part in the procedural inflexibility experienced at COP 6 (part I). The fact that Nigeria, an OPEC country,[4] held the post of G-77 Chair no doubt contributed to the G-77's reticence at allowing procedural relaxation or innovation at that session. However, growing frustration among developing countries (and indeed the Russian Federation) at the erosion of procedural equity and transparency safeguards also exerted an important influence. Discontent at the exclusive friends of the Chair negotiations during COP 4 (see Chapter 9), along with the multiplication of English-only informal groups throughout the post-Kyoto negotiations, and indeed the procedural breaches in Kyoto, led many developing countries to genuinely fear that rules for the conduct of business would not be sufficiently respected for the finale of the post-Kyoto negotiations at COP 6.

It is important for the organizers of the negotiation process to be adept at distinguishing legitimate procedural concerns from illegitimate filibustering. This can be very difficult. At COP 6, for example, COP President Pronk arguably gave too much credence to procedural objections coming from the G-77 Chair, and the consequent delay in taking organizational decisions led to damaging uncertainty. President Pronk undoubtedly feared, however, that disregarding G-77 concerns might antagonize the Group.

An important factor in countering procedural opportunism is for the organizers of the negotiation process to ensure that the most visible and highly regarded procedural rules (e.g. provision of interpretation, making documents available as promptly as possible, scheduling no more than two meetings at any one time) are adhered to as much as possible. Beyond such caution, the presiding officer needs to exercise firmness in responding to clear cases of procedural obstruction, and careful judgment in distinguishing when a concern is legitimately and genuinely held, and when it constitutes a delaying tactic. The ability to be firm in a diplomatic and politically acceptable way, perhaps citing precedents or even using humour, is one of the hallmarks of an effective Chair (see Chapter 4).

Summary and concluding remarks

This chapter has shown how the rules and informal practices for the conduct of business are focused chiefly on promoting procedural equity and transparency and, because of this, they can also imply high transaction costs. In order to increase the efficiency of the climate change negotiations, the rules are therefore applied pragmatically, and often also bypassed, through the use of informal mechanisms to which the rules do not apply (notably informal negotiating groups and informal texts). Especially in the latter stages of important negotiating rounds, it can also be tempting for the organizers of the negotiation process to seek to relax some of the rules in the face of time pressure. Developing countries typically place much more importance on upholding the rules for the conduct of business than industrialized countries, especially as regards interpretation and scheduling of meetings. This is due to their generally greater need for the procedural equity and transparency safeguarded by the rules. It is important, however, not to go too far, so that the negotiating process is rendered illegitimate. The extent to which parties are prepared to accept the relaxation of rules for the conduct of business depends very much on political will. This, in turn, depends partly on the extent to which the rules have been broadly respected throughout the negotiating round as a whole. Another dimension to the rules for the conduct of business is the way in which these can be used for obstructionist purposes, thereby reducing the efficiency of the negotiations. In this regard, it is important that the organizers are able to address such procedural opportunism through firmness and by minimizing opportunities for procedural challenges.

Decision-making Rules

The overwhelming majority of the parties is willing to adopt an international instrument... That majority should not be frustrated (Estrada, 1997, report to COP 3).

Introduction

Decision-making rules are central to the organization of negotiations in intergovernmental regimes, as it is through these rules that the regime must sustain the delicate balance between state sovereignty and global interests upon which it is based. As discussed in Chapter 3, however, the climate change regime does not have an agreed rule for taking most substantive decisions, due to the inability of the parties to agree on voting rules in the rules of procedure. Although the remaining rules of procedure are routinely applied at each session, rule 42 on voting remains in contention and is not applied. In its absence, with the exception of decisions for which voting rules are set out in the Convention (adoption of amendments, annexes) all decisions are taken by consensus (Yamin and Depledge, 2004). In this chapter, we explore the implications of the absence of a voting rule and the resulting consensus imperative in the climate change regime.

Consensus and its contested meaning

As Werksman notes, 'what voting rules should operate in the vacuum left by Rule 42 has been the subject of intense debate and speculation'. While 'a technical solution to this deadlock might lie in the rules of customary international law, most delegates seemed to concede that, in the absence of a specified majority voting rule, decisions would have to be taken *by consensus*' (Werksman, 1999, p6, emphasis added). Consensus is indeed widely used as a decision-making rule in the UN system, particularly in multilateral treaties, given that 'states generally eschew the open confrontation that can come with voting' (Werksman, 1999, p7), even when majority rules are in place.

Despite its pervasiveness, consensus, whose literal meaning equates with 'common feeling' or 'concurrence of feelings' (Schermers and Blokker, 1995, p772), is 'a rather elusive decision-making process' (Evensen, 1989, p78), whose operational meaning is not defined in the Convention (Yamin and Depledge, 2004). Most theorists and practitioners agree, however, that consensus is distinct from unanimity, and is generally defined negatively to mean that there *are no stated or formal objections to a decision* (e.g. Yefimov, 1989; Schermers and Blokker, 1995; Werksman, 1999; Yamin and Depledge, 2004). This 'enables parties to acquiesce in the outcome of a decision without having to express open agreement or disagreement' (Széll, 1996, p211), in other words, a party could reluctantly consent to a decision, but then ask for its concerns to be formally noted (e.g. in the report on the session) after the decision is adopted. The converse of this, however, is that a small group, or arguably even a single party, could formally state that there was no consensus on a particular decision, thus potentially preventing the decision from being adopted (Yamin and Depledge, 2004). The qualified nature of the latter point demonstrates the ambiguity and fluidity of consensus; can the formal objection of a single party block a decision? If not, how many objections does it take for there to be no consensus? One interviewee queried, 'if you are adopting something by consensus, what does that mean? Does that mean there are no objections? Does it mean only one objects? If one country pipes up, does it mean you have no consensus? It's not 100 per cent clear.'

In practice, it is extremely rare for a single party to block a decision, as most parties are very reluctant to prevent their allies – on whose support they may depend in other multilateral arenas – from taking action. If a party does not have the support of its negotiating coalition, it is unlikely to isolate itself by blocking a decision. A good illustration of this occurred at SBSTA 16, where Saudi Arabia fought strongly to delete a mandate to hold an inter-sessional workshop on LULUCF. However, its demands were openly opposed by the G-77 spokesperson and more than 20 parties, including other developing countries. The Chair was thus able to declare consensus on holding the inter-sessional meeting, without formal objection by Saudi Arabia. At a more general level, the fact that the US, having repudiated the Kyoto Protocol, nevertheless stated that it 'would not stop others from moving ahead'[1] illustrates the disinclination of parties to reap international opprobrium by frustrating widespread consensus, even if they are procedurally able, and substantively eager, to do so.

The ambiguity of consensus endows the presiding officer with 'considerable discretion to assess whether a party is registering a formal objection, or some lesser level of discontent that will allow a decision to go forward' (Werksman, 1999, p7). Ultimately, however, the presiding officer can only declare consensus with the acquiescence of the parties, and the meaning of consensus can thus became an object of struggle. This was certainly the case during the Kyoto Protocol negotiations, as Chair Estrada, obstructionist parties and others all sought to shape the meaning of consensus for their own strategic purposes.[2]

However defined, consensus is typically regarded as the decision-making rule of choice in multilateral negotiations (Yamin and Depledge, 2004), and particularly so in global environmental negotiations, where the committed participation of all states is viewed as critical. This was reflected in many interview responses. One interviewee remarked: 'This is the practice within the environmental treaties, you do everything by consensus, even when you have rules of procedure. I really don't know what would happen if we put anything to the vote...people try not to do that.'

A number of interviewees emphasized the potential problems associated with a voting rule in the context of a global issue requiring the engagement of all countries. For one interviewee, 'once you make the leap to a global regime, I don't think you can then exclude key interests through the voting rules. You have to make the regime inclusive.' Several interviewees expressed concern that voting would impact on the legitimacy of the agreement and alienate those parties that lost the vote, meaning that they would be less likely to ratify and implement it. Such views are also shared in the literature. Széll, for example, also a former senior delegate in the climate change regime, notes that 'parties are more likely to respect a decision if they subscribe to its terms than if they are driven reluctantly into observance by means of a majority decision' (Széll, 1996, p213). It is interesting to note that some regimes, mostly older ones, do operate on the basis of regular voting, notably the 1973 Convention on International Trade in Endangered Species. In the case of more recent regimes, however, including the ozone regime and the three Rio Conventions, treaty bodies have preferred to retain the support of all their parties when taking decisions, in order to keep all involved countries on board rather than create divisions and resentments that would mitigate against a cooperative, international response.

Notwithstanding the desirability of consensus decision-making, provisions for majority voting 'if all efforts at consensus have been exhausted and no agreement reached' are common in multilateral environmental agreements. Indeed, such a clause is part of decision-making rules for the adoption of amendments and annexes to the Convention. Placing emphasis on consensus while providing for last resort majority voting in this way aims to balance out the concerns of majorities and minorities. As Széll (1996, p212) explains:

> Such formulations, by specifying a majority vote as a last resort for questions of substance help those parties concerned about the potential of consensus to enable important progress to be blocked by just one dissenting party, whilst the requirement to make every effort to reach consensus gives comfort to those parties which dread a decision being taken against their deeply held views.

Last resort majority voting, however, is precluded for almost all decisions under the regime, so that decisions can *only* be taken by consensus. Although the absence of last resort majority voting and the consequent consensus imperative is now widely taken for granted in the climate change regime, it has nevertheless had an important impact on negotiations throughout the regime's history.

The impact of the consensus imperative

The absence of a voting rule and the consequent consensus imperative were most keenly felt during the Kyoto Protocol negotiations. For one interviewee, 'it hung over the process like a cloud...the dynamics of the AGBM would have been different with a voting rule in place'.

Greater procedural equity?

At one level, the impact of this consensus imperative could be said to have generated *greater procedural equity* in the negotiations, as it meant that even the views of small minorities had to be accommodated in a consensus, rather than simply being outvoted. As one interviewee argued, 'if the case deserves attention and they [minorities] could be steamrollered, you would want them to have the opportunity to reach a decision by consensus'. The resulting content of the Protocol therefore undoubtedly accommodated a greater range of positions than would otherwise have been the case with a voting rule in place. One interviewee summarized this effect as follows:

> You have to buy them [small minorities] in...politically or economically you have to buy them in, which is a different dynamic, and that's the dynamic that's been followed in the climate change regime. Negotiated buy-in...that's the logic of no majority voting rules.

The best example of such negotiated buy-in is that of the OPEC countries and their specific proposal, later supported by the G-77 and China, for a fund to compensate them for potential economic losses due to climate change mitigation action. This proposal was strongly and universally opposed by Annex I parties. However, because of the imperative need for consensus, the Annex I parties did engage in negotiations on it, eventually agreeing a text that all could live with. As Estrada stated in his interview, 'we got the agreement by consensus, simply by offering the oil producers some paragraph in article 2 and another paragraph in article 3 [on possible adverse effects of mitigation action]...'. Without the implicit threat that OPEC would block a consensus, it is unlikely that the Annex I parties would have agreed to the inclusion of these clauses in the Protocol, or even talked about the issue. Moreover, there is evidence that side payments were also made to OPEC countries outside of the formal negotiation process. Oberthür and Ott, for example, reveal that Japan promised to review its oil imports, establish stronger bilateral contacts and take steps to boost Japanese investment in Saudi Arabia in return for that country's cooperation in Kyoto (Oberthür and Ott, 1999).

Another example is that of Australia, whose demand for a significant emissions *growth* target in the Kyoto Protocol was almost unique among Annex I parties and not viewed with sympathy.[3] However, Australia succeeded in achieving a +8 per cent target, in the face of rumours that it was otherwise prepared to walk away from the negotiations (although not necessarily to block consensus). Moreover, on the last night of negotiations, Australia secured a clause in the Protocol relating to the treatment of carbon sinks that applies almost exclusively to that country to considerably lighten the effort required to meet its target. There is little doubt that other parties who might have objected did not query this clause for fear of upsetting a consensus. On this issue, one interviewee stated, 'I don't think it [Australia] would have ever asked [for this clause] if there had been a two thirds majority vote'.

Examples of such negotiated buy-ins are legion in the climate change regime, and have continued in the post-Kyoto period. For example, considerable efforts are repeatedly exerted to secure the cooperation of OPEC countries. It is not uncommon for OPEC countries, notably Saudi Arabia, to threaten to block negotiations on other agenda items, if issues of concern to them do not advance sufficiently rapidly. It is undoubtedly the case that, without the threat of blockage on the part of a handful of OPEC countries, the issue of adverse impacts of mitigation measures would have fallen off the climate change agenda a long time ago. In a similar case to that of Australia mentioned above, the need for consensus meant that the Russian Federation was also able to exert sufficient power to secure special consideration for itself in the Marrakesh Accords, altering a decision agreed at COP 6 (part II) to grant it increased credits from forest management activities (see decisions 11/CP.7 and 12/CP.7). In an only slightly less blatant exercise of political muscle, Canada and Japan were similarly able to secure very large amounts of credits for themselves for the same forest management activities at COP 6 (part II). These special allowances were resented by many other parties, including developing countries, and, had they gone to a vote, may not have been agreed. As one interviewee put it, 'the need to give small gifts to everybody would have been reduced with a voting rule'.

The foregoing examples suggest that greater procedural equity, achieved by the consensus imperative, gives a disproportionate advantage to the more laggard parties, thus shifting the resulting agreement towards an environmentally weaker substantive content, what might be called the 'law of the least ambitious' (Wettestad, 1999, p25). A consensus decision-making rule does indeed tend to lead to 'least common denominator solutions' (Werksman, 1996, p60, see also Yamin and Depledge, 2004), as it is parties who advocate the weakest substantive effectiveness that are most prepared to break the consensus. The minority view of AOSIS, for example, advocating a much stronger target, did not enjoy similar leverage to that of OPEC or Australia during the Kyoto Protocol negotiations.

Raising transaction costs and procedural blockage

On a day-to-day basis, the consensus imperative *raises the transaction costs* of the negotiation process, as greater effort is required to secure even minor decisions. One interviewee commented, 'it's very frustrating to experience the lack of voting rules and not being able to get firm decisions before you really have to', while another remarked, 'our inability to take any votes...makes life difficult, because...you have to deliberate so intensely over minutiae because of the wishes of a small group'.

Moreover, a consensus requirement can allow a handful of parties to hold the vast majority hostage. There was certainly the fear in the Kyoto Protocol negotiations that, at the moment of adoption of the Protocol, a small group of obstructionist parties, or even an individual delegation, would declare that there was no consensus and prevent the decision being taken. An interviewee described the threat of procedural blockage in the following terms:

> It definitely was a big issue in the corridors...What are we possibly going to do when Saudi Arabia and Kuwait raise their flags. At the end of the day it didn't matter...But the stakes had never been as high as they were in Kyoto, so there was legitimate reason to suspect that, well, this really will be where OPEC draws the line in the sand, and says 'we're willing to put up with the anger of the international community that it was clearly us that blocked the decision, because the decision has such a potential impact on our interests'. But at the end of the day, they didn't.

The threat of procedural blockage to the overall package of negotiating issues was much less of a concern during the post-Kyoto negotiations. The fact that the OPEC countries had not blocked the adoption of the Kyoto Protocol had implicitly signalled that this most obstructionist group of parties would continue to pursue their strategies within the regime, rather than through its overall blockage. The adoption of clauses in the Kyoto Protocol on the adverse effects of mitigation measures provided a vehicle for them to do so very effectively. Other parties with very strong positions on certain issues in the post-Kyoto negotiations – notably Australia, Canada, Japan and the Russian Federation – were deemed unlikely to go so far as to block a consensus. However, there was still widespread concern among the organizers of the negotiation process that some of these individual parties might refuse to join – as opposed to actively obstructing – a consensus, which in itself would be severely damaging to the prospects for entry into force and future development of the Kyoto Protocol, and indeed confidence in the regime. Such individual withdrawal, even if it allowed other parties to go ahead, would therefore, in effect, also constitute de facto procedural blockage.

Overcoming the threat of procedural blockage

Working to overcome the threat of procedural blockage is therefore a critical task for the organizers of the negotiation process. As discussed above, the main strategy for doing so is simply to encourage parties to strive extra hard to achieve consensus, through negotiated buy-in and otherwise exchanging concessions. This, indeed, is the essence of negotiation. There are other procedural tools, however, that the organizers of the negotiation process can wield to help secure a consensus when the threat of procedural blockage hangs in the air.

Use of safety valves

Procedural safety valves (see Werksman, 1999) can be very helpful in enabling parties to relinquish their demands, yet still save face or secure a guarantee that their interests will not be ignored. The most common safety valve is for a reluctant party to make an explanatory statement on adoption of a decision, setting out that party's own views and interpretation of the decision. For added effect, the party may ask for the statement to be noted in the report on the session (Yamin and Depledge, 2004). There are many examples of this in the climate change regime, and indeed the trend of requesting explanatory statements to be recorded in reports appears to be on the rise. At COP 6 (part II), for example, on adoption of the Bonn Agreements, the US sought to 'emphasize that our not blocking consensus... does not change our view that the Protocol is not sound policy', listing areas of particular concern, notably its understanding of decisions relating to financial commitments (FCCC/CP/2001/MISC.4). The US did so again at COP 7 (FCCC/CP/2001/MISC.9). Also at COP 7, the Republic of Korea made a statement on its understanding of the status of unilateral projects under the just-adopted decision on the CDM (COP 7 report part I, paragraph 103). At SBSTA 16, New Zealand, revealingly stressing that it was 'not a country to block', followed up the adoption of conclusions on the IPCC TAR by expressing regret at both their content and the process of their negotiation (SBSTA 16 report, paragraph 14). SBSTA 19 saw a spate of explanatory statements, with four parties recording their own interpretations of a draft decision recommended for adoption by the COP on LULUCF projects under the CDM, and the Russian Federation expressing concerns over the adopted conclusions on the scientific, technical and socio-economic aspects of mitigation (SBSTA 19 report, paragraphs 10 and 20).

Another safety valve open to reluctant parties is to obtain an undertaking that issues of importance to them will be taken up at a future session, or at least placed on the provisional agenda (see Werksman, 1999). This can help to secure support back home for decisions of the climate change regime, as domestic constituencies can be reassured that their concerns are on the table at the international level. Two good examples here include the case of Iceland, which was able to agree to the adoption of the Kyoto Protocol, despite admitting that its (large growth) target was not achievable, by securing an undertaking that the issue of the 'impact of single projects' would be taken up at COP 4 (decision 1/CP.3, paragraph 5d).[4] A second example is that of Canada, whose ability to join in with the consensus on the Bonn Agreements at COP 6 (part II) was greatly facilitated by the (eventual and reluctant) acceptance of the parties that its proposal concerning credits for the use of cleaner energy would be subject to future discussion (see COP 6 (part II) report, part I, paragraphs 64 to 67 and part II, paragraph 1).[5]

The increasing use of such safety valves, while allowing decisions to be adopted, raises the danger that contentious issues will simply continue to fester, and may also generate problems with implementation, where countries have different understandings of what has been agreed.

Forcing consensus

Another means of overcoming the threat of procedural blockage is for the Chair to exploit the leverage granted to him/her by the lack of an agreed definition of consensus. As discussed in Chapter 4, this is an extremely delicate task, requiring the Chair to make a careful judgment on whether, when pushed, a party will consent to a decision, or whether the issue is of fundamental importance to that party, and it will therefore insist on obstructing a consensus. In this respect, the Chair relies on the reluctance of parties to be seen to actually block a decision that the other parties agree to (Yamin and Depledge, 2004). Such forcing of consensus is relatively rare, but has been used to good effect on a number of occasions.

Chair Estrada, during his tenure as Chair of the Kyoto Protocol negotiations, undoubtedly made the greatest use of the Chair's prerogative to declare consensus, often taking considerable risks in doing so. The most extreme example occurred at AGBM 8 in 1997 (the last negotiating session before Kyoto) where, after protracted debate over the wording of a relatively minor piece of text, Chair Estrada declared that there was consensus to adopt one of the options, although three parties (Australia, Canada, US) had argued against it. Estrada stated that, in order to overturn his ruling, there would need to be a two thirds majority vote of parties (AGBM 8, 1997d). In doing so, he was appealing to draft rules 42.2 and 42.3, which allow a decision on procedural matters to be taken by a vote. The presiding officer is empowered to rule on whether an issue is procedural or substantive, and that ruling can only be overturned by a majority vote. The extent to which Chair Estrada could rely on these unadopted *and unapplied* provisions to force a vote, however, was very dubious (Werksman, 1999). Many delegates intervened to urge him not to hold a vote. Estrada eventually desisted, but stated that, in doing so, he understood his ruling was no longer under challenge, and pushed through his interpretation of consensus. The debate that led to this outcome (see Box 8.1), usefully illustrates several points raised in this chapter, including contrasting definitions of consensus, the reluctance of parties to resort to a vote, the implicit threat of some OPEC states to block consensus in Kyoto, and Estrada's resolve to force decisions through.

While the text pushed through at AGBM 8 was of little practical consequence and was soon superseded, Estrada later secured agreement at COP 3 on various articles of the Kyoto Protocol through similarly forceful declarations of consensus. Estrada declared that there was consensus to adopt Article 10 on general commitments, Article 12 on the CDM and Article 17 on emissions trading, despite open expressions of dissatisfaction with those articles by a significant number of parties. He declared consensus to delete the former draft article on voluntary commitments in similar circumstances. In doing so, he surmised that, while parties would inevitably stick to their positions until the very last minute, when faced with the prospect of actually blocking the Protocol, they would bow to his declared consensus. He was proved right. Interestingly however, the Kyoto Protocol *as a whole* was agreed in the Committee of the Whole through a very clear and unambiguous

Box 8.1   The 'almost vote'

Estrada:	... if there are no other comments... I will rule that *there is consensus* to keep that text *with the exception of three countries* which are not agreed on that, and we keep that text with consensus. It's so decided. [gavel]
US:	I do not accept your definition of consensus... *consensus means lack of a stated objection*, and I have very clearly stated my objection, as have others.
Venezuela:	*There can be no general consensus if any single country, and far less if three countries, raise objections.* We are ready to contradict your ruling.
Estrada:	OK. I have ruled that we have consensus in a situation where we have three countries with a different view. This point has been challenged by... Venezuela. In order to overcome my ruling, he needs two-thirds of the votes... I will call a roll-call... that will clarify the rest of the work for us.
Egypt:	Legally speaking... you have a correct ruling... What others were speaking about was *unanimity*... but your ruling means we have consensus minus, which means you reflected the *sense of the negotiating process*... I appeal to everybody to *go ahead with your ruling* without... voting which would *create a new precedent.*
Mauritania:	We all cherish your authority... We are preparing a new legal agreement, we all want it to be *ratified by every country.*
Hungary:	There is no need for any general consensus. You were very good in the past, and you will be very good even now, at understanding the *general feeling* in the room.
Saudi Arabia:	There is *no need to use your hammer*... we ask you to relax!
Estrada:	I understand the challenge to my ruling has been withdrawn, so we don't need to go to the vote.
Kuwait:	Consensus, it is a very important matter. It will face us in Kyoto... I want an *official United Nations legal definition of consensus.*
Estrada:	From the very beginning... a group of countries was trying to stop the process... *I will do everything to overcome those countries*... I am not going to be [held] hostage.

AGBM 8 (1997d); extracts, emphases added

expression of consensus by all the parties – indeed, there were no objections when Estrada declared its *unanimous* approval.

Other cases of forcing through consensus have punctuated the history of the climate change regime. At a late night meeting of SBSTA 9, for example (held in conjunction with COP 4 in 1998), the US objected to recommending a draft decision on technology transfer for adoption by the COP (see ENB, 1998). This followed protracted small group consultations, along with a compromise proposal by the SBSTA Chair, Chow Kok Kee. Chair Chow, aware that the US was the only objector and that failure to accept the draft decision would provoke ill will among developing countries, pushed through the acceptance of the text in the SBSTA, and said the US position would be noted in the report. The US was unhappy with this safety valve solution, and vowed to raise the issue in the COP. Interestingly, however, the US did not then object to the formal adoption of the decision by the COP less than 36 hours later. This suggests that the SBSTA Chair was indeed right to push the decision through, as US concerns were not founded sufficiently deeply to cause that country to take the politically difficult step of preventing the actual adoption of the decision in the COP itself.[6]

There are other cases where Chairs have declared consensus after clear expressions of dissent, or have even threatened to call for a vote, where they surmised that objections were not founded on fundamental concerns. This occurred, for example, at SBSTA 14 in 2001, where Chair Dovland eventually overrode objections by a handful of mostly OPEC parties to the adoption of conclusions on the issue of policies and measures (see ENB, 2001, p9). The seemingly rather trivial question was whether the SBSTA could decide on terms of reference for a workshop before the COP had taken a formal decision on the broader issue of policies and measures, which, due to the failure of COP 6, it would not be able to do until after the scheduled dates of the workshop. After threatening to call a vote on what he saw as a procedural matter, Dovland ruled that the SBSTA was entitled to decide on the workshop, following an earlier COP mandate. He was able to take such strong action as he judged that the objections were procedural in nature and motivated by filibustering rather than fundamental national interests, and he had the clear support of the vast majority of parties, including explicit support from other developing countries, to adopt the conclusions.

Forcing consensus is always a very risky strategy. If a Chair misjudges the situation, and declares consensus when in fact the party concerned is not prepared to acquiesce, then the resulting decision risks illegitimacy. Although this has not happened in the climate change regime, there are cases in the broader international arena where decisions pushed through in the face of a clear objection have then been disowned as illegitimate by the objecting party (or parties) and have triggered serious disquiet over the decision-making process on the part of others. A high-profile case occurred at the sixth COP of the CBD in 2002, where Australia raised a formal objection to certain elements in a draft decision on alien species presented to the COP plenary, and did not agree to simply recording its position in the report (see CBD, 2002, paragraphs 294–324). After inconclusive consultations behind the scenes, the President nevertheless adopted the decision as presented to the COP. Prior to doing so, the President made clear her view that 'consensus did not mean unanimity, but, rather, broad agreement'. As well as Australia's formal objection, this led to protests over the decision-making procedure from other parties, and lingering doubts over the legitimacy and legality of the decision.

Alternatives to adoption

Another strategy employed by the organizers of the negotiation process to overcome procedural blockage has been to use alternatives to adoption in order to register agreement among the vast majority of parties. This strategy has proved useful where a sufficient number of parties were indeed prepared to block adoption, so that the Chair was unable to force through consensus.

The most high-profile example here is the Geneva Ministerial Declaration (see COP 2 report, part II), which was negotiated by ministers at COP 2 in 1996. The aim behind drafting such a declaration, mooted by several parties and the organizers, was to give impetus to the Kyoto Protocol negotiations. The final version was a strong text, endorsing the findings of the IPCC's Second Assessment Report and calling for the adoption of legally binding emission targets at COP 3.

The negotiations on the Declaration made it clear that there would be a 'trade-off between the strength of the content and the number of supporters' (Oberthür, 1996, p199). The strength of the Declaration was too great for a minority of parties, namely, 13 OPEC states, the Russian Federation and Australia[7], who raised formal objections to it. However, the majority of parties were not prepared to weaken the Declaration to try to muster a consensus, but instead sought to move ahead and give it recognition in whatever way possible. Therefore, on advice of the secretariat, COP 2 President Chimutengwende proposed that the Declaration be *taken note of* and annexed to the report of the session, with objections (and indeed statements of support) also recorded in the report (see COP 2 report, part I, section IV).[8] This proposal met with 'sustained applause of the great majority of delegations' (Oberthür, 1996, p200), which undoubtedly influenced the opposing parties' decision to acquiesce to this proposal. Although the Declaration was not adopted, the fact that formal recognition of it was achieved despite the objections of a small group of parties, and that the identity of these parties was clearly revealed, was critical to underscoring the will of the majority of parties not to let a small group prevent progress. As one interviewee put it:

> The declaration was quite important ... because having a footnote of the parties that didn't support it was such a clear indication of the special interests that were trying to block the process and the rest of the world. It was a really important moment when actually they allowed ... the declaration to move forwards and it became the basis for the beginning of a legally binding target.

The G-77 and China later 'expressed concern with the procedure used for adoption of the Geneva Declaration' (ENB, 1996, p12), but stopped short of asking for this concern to be formally noted or recorded in any way, suggesting that the Group, as a whole, tacitly supported it. Taking note of the Declaration in this way was thus as significant, in fact perhaps more so, than if it had been formally adopted. Moreover, the fact that the Declaration was not formally adopted is now almost forgotten; it was the fact that agreement was registered that mattered.

A second example of an alternative to adoption occurred at SBSTA 3, during the same sessional period as COP 2. The SBSTA was experiencing extremely laborious and frustrating negotiations on the IPCC's Second Assessment Report, where a small number of mainly OPEC parties were objecting to a strong endorsement of the report's findings, preferring to highlight scientific uncertainties. In frustration, SBSTA Chair Faragó called for a 'show of hands' of those who opposed the text he had put forward as a compromise. The result indicated that 11 countries – all OPEC members plus China – did not agree with it, and the majority of delegates did (ECO, 1996). While this 'show of hands' did not, of course, constitute a formal vote and nor did it represent a formal decision, its impact was the same in registering the majority view on the issue. Such a strategy of using a 'show of hands' – in effect an informal vote – has only very rarely been used in the climate change regime, but does have the potential of helping to highlight the wishes of a large majority, where consensus is being blocked.

Summary and concluding remarks

This chapter has explored the implications of the consensus imperative that exists in the climate change regime, along with strategies for overcoming the resulting threat of procedural blockage. The experience of the Kyoto Protocol negotiations, in particular, demonstrates that the net effect of the consensus imperative is ambiguous. It bolsters procedural equity by ensuring that the views of small minorities are taken into account, thereby securing the greater inclusivity and legitimacy of the final agreement. However, the consensus imperative combined with the presence of some parties who would be happy to see the negotiations fail can have the effect of weakening the substantive strength of the eventual agreement while raising the transaction costs of the negotiations. Moreover, it opens up the possibility of veto by a small minority of obstructionist states that, in itself, would be highly procedurally inequitable.

That the consensus imperative did not lead to the vetoing of the Protocol was due, at least in part, to the mitigating actions of the organizers of the negotiation process, notably Chair Estrada. Indeed, the lack of a voting rule and an unclear definition of consensus in the context of, as one interviewee put it, 'a very tough Chairman with creative definitions of consensus' may even have been an advantage to reaching agreement. Estrada was able to cleverly adapt the consensus requirement 'to the various decision challenges at hand' (Wettestad, 1999, p216), introducing an element of flexibility into decision-making that was critical given the complexities and sensitivities of the negotiations. Several interviewees agreed that the absence of formal decision-making rules may have been 'a bit of a blessing in disguise'.

Perhaps surprisingly, the absence of a voting rule, and the resulting consensus imperative, is no longer seen as a major concern in the climate change regime. Indeed, the non-adoption of the rules of procedure more generally is viewed as little more than a background irritation. This is evidenced by the fact that consultations by COP Presidents on the rules of procedure have almost ground to a halt, with no serious attempts to resolve the issue since COP 3. The fact that the Kyoto Protocol was adopted by consensus, against expectations, may have contributed to the downplaying of the importance of agreeing a voting rule. The potential advantages of the flexibilities of consensus, however, were only realized in the case of the Kyoto Protocol negotiations because of the presence of a strong presiding officer with good judgement, a situation that cannot always be guaranteed in multilateral negotiations.

Taking decisions on the basis of consensus will almost always be preferable in an intergovernmental setting, and a voting rule would certainly not provide a panacea to the climate change regime. Where countries' fundamental interests are at stake, out-voting them is unlikely to be conducive to the continuation of a legitimate, inclusive, almost universal regime. However, the number of issues on the agenda of the regime bodies is multiplying, together with the technical and rather detailed content of the decisions needed to advance those issues. The need for consensus on each and every action taken under the regime certainly threatens to delay the process, not least by opening up opportunities for deliberate filibustering. In this context, the benefits in terms of increased efficiency of using voting on more detailed matters might outweigh the small losses in terms of procedural equity. In this respect, it is interesting to note that the three limited membership bodies set up under the Kyoto Protocol – the CDM Executive Board, the Article 6 Supervisory Committee and the Compliance Committee – all provide for last resort majority voting for their decisions (Yamin and Depledge, 2004). Given the political fractures that currently exist within the climate change regime, along with the continued influence of obstructionist parties, it would be unwise to place too much faith in (the ambiguity of) consensus and to be complacent over the absence of an agreed voting rule.

Negotiating Arenas

... one of the central organizational problems [is] almost always to strike ... a balance between a large, formal 'debating society' in which all sovereign states are allowed to participate on a full and equal basis, and a smaller, informal setup, where most of the collective learning process goes on, where solutions are tested and finally agreed on before being referred to the plenary (Freymond, 1991, p130).

Introduction

Like all complex multilateral processes, the climate change negotiations play out in a variety of arenas, from the formality and ceremony of the COP plenary, to the chaotic frenzy of behind the scenes deal-making. The choice of arenas throughout a negotiation, and how these arenas are managed both individually and collectively, is a key dimension to the organization of the negotiation process. In this respect, a dilemma common to complex multilateral negotiations, and a recurring theme throughout this chapter, is how to balance the efficiency gains of small, informal, closed groups against the transparency and procedural equity provided by large, formal, open arenas. There are no set rules governing the nature, structure and proceedings of *informal* groups – indeed, these are informal by virtue of having abandoned some, but not necessarily all, of the formal rules for the conduct of business (see Chapter 7) – thereby opening up considerable room for improvisation in their design. The organizers of the negotiation process have taken advantage of this flexibility, improvising with diverse types of informal groups (and formal/informal hybrids), and combining these in various ways with formal settings at different stages of negotiating rounds. This chapter explores the characteristics of the main types of negotiating arenas used in the climate change regime (summarized in Table 9.1), and discusses how these arenas have been used at varying points in the history of the climate change regime to seek to promote productive negotiations.

This chapter begins with some observations on how the overall institutional setting of the negotiations has varied for the Kyoto, post-Kyoto and post-Marrakesh rounds. We then examine formal, open settings, including COP and subsidiary body plenary meetings, along with the hybrid formal/informal groups that often serve as working bodies for the COP. We then turn to informal groups that are open to all parties, including contact groups and informal consultations, before focusing on 'friends' groups and associated 'shuttle diplomacy', that is, informal meetings convened personally by the presiding officer, whose participation is limited to invited parties. The chapter then moves briefly into the corridors to explore the unofficial talks behind the scenes which, although unregulated by the official process, are often the source of the final deal in a negotiation.

Table 9.1 *Main types of negotiating arenas in the climate change regime*

Category of arena	Negotiating arena in this category
Formal, open	Plenary meetings (COP and subsidiary body)
	COP Working Bodies
Informal, open	Informal groups (e.g. contact groups, negotiating groups)
	Informal consultations
Informal, closed	'Friends' groups
Unofficial	Behind the scenes

The climate change institutions

The institutional structure of the climate change regime was introduced in Chapter 3. To recap, the COP serves as the 'supreme' decision-making body, assisted by two subsidiary bodies, one on Implementation, and one on Scientific and Technological Advice. The SBI and SBSTA are more technical bodies that usually meet twice-yearly to prepare decisions for the high-profile annual COP.

A separate body was convened to conduct the Kyoto Protocol negotiations, known as the Ad Hoc Group on the Berlin Mandate (AGBM). The conduct of negotiations in a specially convened, dedicated body rather than using the existing subsidiary bodies had several advantages. Importantly, it provided a clear channel for the exercise of strong process-oriented leadership by that body's Chair, Raúl Estrada, enabling him to take responsibility for the full negotiation process in a coherent manner (see also Chapter 4). According to an interviewee, 'you need that kind of single unitarian leadership to push that sort of thing through, where everyone feels a common kind of commitment'. A single body also conferred greater status and distinction on the process, providing a focus for public and media attention, and ensuring that more time was dedicated to the negotiations. Another advantage was that dealing with all issues under one institutional umbrella helped to maintain coherence in the negotiation process, while a sense of community and identity built up around participants. It also helped to insulate the Protocol negotiations from controversial debates taking place within the permanent subsidiary bodies, such as the protracted debate over the IPCC's Second Assessment Report, which was contained within the SBSTA. Commenting on the decision to set up the AGBM in 1995, Oberthür and Ott noted that 'establishing a special ad hoc body should … improve the prospects for quick progress over the next two years to meet the deadline of 1997' (Oberthür and Ott, 1995, p146).

In contrast, the post-Kyoto negotiations were conducted within the existing two subsidiary bodies, with the issues under negotiation being allocated to one or the other body. Issues that cut across the mandates of the SBSTA and SBI – the flexibility mechanisms, adverse effects, and later also capacity-building – were jointly allocated to both bodies. The issue of compliance, for its part, was addressed in a joint working group on compliance established under the subsidiary bodies by COP 4. Therefore, despite covering a package of issues enshrined in the Buenos Aires Plan of Action (BAPA), the post-Kyoto negotiations were much more disaggregated than the previous Kyoto negotiating round in terms of their institutional structure and presiding officers. Although joint meetings of the subsidiary bodies on their joint issues ensured that at least those issues were dealt with by one body, there was no sense of unitary leadership encompassing the whole BAPA (see also Chapter 4).

Since the adoption of the Marrakesh Accords and the close of the post-Kyoto negotiating round, negotiations have continued in the two subsidiary bodies. This, however, is entirely appropriate, given that these are routine regime development negotiations, without any packaging element to them.

With these observations on institutional structure in mind, we now turn to the various arenas in which the climate change negotiations play out.

Formal open arenas: plenary meetings and variants

Formal plenary meetings

The central stages for the climate change negotiations are the plenary meetings of the COP and subsidiary bodies, which are typically attended by all parties present at the session and are open to non-state observers, including the media. These plenary meetings are required to adhere to all formal rules and established practices for the conduct of business, such as seating arrangements in alphabetical order,[1] provision of interpretation, rules governing the right to speak and quorum requirements, and therefore enjoy maximum transparency and procedural equity (see Chapter 7). The presence of full transparency and procedural equity safeguards means that plenary meetings are the *only arenas where formal decisions can be taken* (Yamin and Depledge, 2004). This includes procedural decisions, such as adoption of the agenda and organization of work. Formal plenary meetings therefore serve as a kind of governing body, deciding on the agenda, establishing informal groups to take up agenda items, and finally adopting conclusions or decisions.

It is taken for granted in the climate change regime, and indeed in other international forums, that plenary meetings are not conducive settings either for active debate or bargaining. The high transaction costs involved in adhering to formal rules for the conduct of business, the large number of parties present, and the admission of the media and NGOs all serve to constrain frank and spontaneous negotiation (Yamin and Depledge, 2004). One interviewee echoed the sentiments of many others by saying: 'If you ask delegates, they all know, we can't negotiate in plenary. Everybody will tell you the same.' Interestingly, some presiding officers have tried, on occasion, to provoke more open debate in a plenary setting by convening an *informal plenary*. Box 9.1 below discusses this apparent oxymoron and its limited, but useful, applications in the climate change negotiations so far.

In the SBSTA and SBI, every item on the agenda will first be addressed in a plenary meeting. This initial plenary debate, however, will almost always be confined to the delivery of prepared statements by delegations and negotiating coalitions, rather than any substantive bargaining. Except for the most uncontroversial ones, agenda items will then be referred to informal groups for further work. Indeed, there are signs that initial plenary debate in the subsidiary bodies is becoming increasingly brief, as parties save their energy and words for the real work that they know will take place in informal groups. Subsequent plenary meetings will then be held to hear reports from the informal groups, take stock of progress in the negotiations, and finally to adopt decisions. In the case of the COP, plenary discussion is even more limited. An initial plenary meeting will usually refer all debate on most of its substantive agenda items to the subsidiary bodies or another working body (see below).

The functions of formal plenary meetings

As noted above, the main purpose of plenary meetings is not to negotiate. Rather, plenary meetings provide all parties with an equal and public opportunity to *posture and stake out their positions*, formally placing their views on the table and making a bid for their preferences in the negotiations. The prepared opening statements of key players – notably the EU, G-77 and China and the US – thus often serve as helpful barometers for the presiding officers and secretariat, and no doubt other parties, to get a feel for the extent of flexibility these players are prepared to show, and therefore how the session is likely to evolve. A public opportunity for a delegation to express itself without interruption with all procedural safeguards in place tends to be particularly important for developing countries, who can feel that they have placed their national positions on record, even if they then have limited influence in the bargaining process (see also Chapter 7).

Delivering prepared statements in an open plenary can similarly serve as an avenue for *letting off steam*; parties make strong interventions, but having expressed themselves publicly then take a more constructive approach in the bargaining process. Parties might deliver such hard-line statements motivated by genuine frustration, the need to appease extreme factions within a delegation or coalition, or to 'play to the gallery' (Fisher et al, 1992, p33) of NGOs and the media. For example, at the closing plenary of AGBM 8 on the eve of COP 3, the US made an unexpected statement that the Protocol should not adversely affect its military capability (USA, 1997). The US delegation had been under pressure from the Pentagon to raise this issue, and had chosen to do so at the closing plenary where it knew the media and NGOs would be present, to reassure the Pentagon that its concerns had been taken on board (personal communication). Having been seen to convey a strong message in a public arena, the US delegation then engaged in bargaining on the issue in informal and unofficial groups in a much more discreet and conciliatory manner.

Once agenda items have been referred to informal groups, formal plenary meetings can perform a very important role as *stock-taking* forums, where the presiding officers of the various informal groups are asked to report on their work, and delegates are given the opportunity to comment. Holding such stock-taking plenary meetings can be critical to mitigating the weaknesses of informal groups (see below), in particular enhancing transparency relative to NGOs and minimizing the extent to which parties not able to participate effectively in the informal groups, especially non-Anglophones, feel included.

Stock-taking plenary meetings can thus function as a system of checks and balances, allowing parties not involved in the informal groups to raise concerns, if necessary, at the emerging results of the groups. A good example of where small, poorly resourced delegations successfully used a plenary in this

---

Box 9.1 Informal plenary meetings

A couple of examples exist in the climate change regime of where *informal* plenary meetings have been convened to encourage more open debate among parties. In these cases, a formal plenary meeting has simply been declared by the presiding officer to now have the status of an informal meeting, while retaining all the trappings – interpretation, sound recordings, alphabetical-order country seating, use of flags to signify desire to speak – of a formal plenary meeting. The informal designation, however, along with a more relaxed chairing style on the part of the presiding officer, is usually sufficient to provoke a somewhat more open debate.

At COP 8, for example, COP President Baalu converted the COP plenary to an informal setting to allow for an open exchange of views on inputs for a draft ministerial declaration. The informality of the setting was useful in providing a safe and politically less controversial forum for parties to candidly express simmering concerns regarding the possible mention of future commitments for non-Annex I parties in the declaration. Containing such heated exchanges in an informal arena, while ensuring openness and transparency through a plenary setting, helped to prevent bitterness over the issue from spilling out into other aspects of the negotiations.

Another case occurred during the early Kyoto Protocol negotiations, where Chair Estrada regularly convened informal plenary meetings of the AGBM as an alternative to establishing informal groups. In doing so, he sought to achieve the best of both worlds, that is, an open, fully transparent forum (unlike informal groups), but one where active debate took place (unlike formal plenary meetings).

The subsidiary bodies also occasionally convened in an informal plenary setting in the pre-Kyoto period. The increase in the burden of work of the subsidiary bodies post-Kyoto, however, means that informal groups and informal consultations are now preferred over the more time-consuming informal plenary.

Note: 1 Another attempt at designating a plenary meeting as informal was the case of the *informal high-level plenary* (IHLP) used at COP 6 and, to a lesser extent, COP 6 (part II). The President's ambitions for this forum, however, were greater, with the IHLP envisaged as a full working body of the COP. It is therefore discussed in more detail under 'COP working bodies' below.

way was in the final stages of negotiations at COP 3 on the issue of how the emerging protocol should address possible impacts of climate change and mitigation measures on developing countries. Here, the informal group Chair reported to a plenary meeting of the Committee of the Whole (a working body of the COP chaired by Estrada – see below) that agreement had almost been reached on a text. However, when that text was circulated in plenary, several LDCs expressed deep concern that it did not make specific mention of their group of countries (see FCCC/TP/2000/2). The ensuing heated debate suggested that certain OPEC countries had enjoyed undue influence over the informal group, and that LDCs had been unable either to participate in it or to make their views heard. The Mauritanian delegate made use of the plenary setting to explicitly draw public attention to the actions of OPEC countries, which had been hidden in the privacy of the informal group (which was closed to observers). The delegate stated, 'I assume that the OPEC countries oppose inclusion of LDCs? Is this correct? If so ... *the international community should know about this*' (CoW, 1997e). Chair Estrada requested the informal group Chair to consult further, and reference to the situation of LDCs was eventually added to the text. In this case, the monitoring and supervisory role of plenary meetings was crucial to ensuring that the views of all were heard.

It is not uncommon for informal groups to fail to reach agreement. In such cases, the negotiating text will be presented back to a plenary meeting of the COP or subsidiary bodies, which will then serve as *negotiating forum of last resort* to resolve the issue at hand. Taking an issue back to plenary in this way – or even just threatening to do so – can be helpful where there is suspicion that representatives of negotiating coalitions are not faithfully reflecting their negotiating mandate, or simply to place pressure on recalcitrant groups to compromise. There are many examples of issues being brought back to plenary, most commonly to the subsidiary bodies. At SBI 19 (held in conjunction with COP 9 in 2003), for example, the Chair of the informal group on issues of specific concern to LDCs reported in plenary that there had been no agreement in the group on a proposed draft text. The coordinator of the LDC negotiating coalition was able clearly and openly to explain his group's problems with the proposed text in plenary and, following debate in plenary, an amendment was found to the satisfaction of all groups.

The presiding officer will, however, often suspend the plenary meeting to convene a small group informal consultation rather than engage in bargaining in the plenary setting, which can be laborious. This approach was followed, for example, for negotiations on the IPCC Third Assessment Report (TAR) at SBSTA 16, where the informal group failed to reach agreement, and had to appeal back to plenary to resolve outstanding disagreement. After a debate in plenary, which usefully exposed the strong but isolated position of the Russian Federation, the SBSTA Chair was able to reach agreement in a small informal consultation, where parties were able to climb down without fear of losing public 'face'.

It is less common for the COP plenary to be used as a forum for last resort bargaining. Due to its more ceremonial and high-profile nature, along with the difficulties faced by most COP Presidents in chairing a spontaneous negotiation (see Chapter 4), parties and the organizers of the negotiation process usually exert considerable effort to ensure that texts presented to the COP plenary for adoption have already been approved in the subsidiary bodies or another working body. There are cases, however, where the COP plenary has served as a negotiating forum. One example was the proposal put forward by Canada on cleaner energy at COP 6 (part II). Informal consultations convened on this issue had not come to a consensus, and an extensive debate took place in plenary before agreement was reached.[2] In this case, open plenary debate was very useful in illustrating and exposing the extent of differing views, in particular, the absence of a united G-77 position. Another incident occurred at COP 9 where neither the informal group nor the convening SBI plenary were able to conclude negotiations on guidance for the LDC fund. In this case, the SBI plenary actually referred the issue back to the COP plenary, an unusual move. This placed even more pressure on parties to compromise, which they eventually did, when the COP President suspended the COP plenary for a small group consultation.

Last minute bargaining in plenary can also occur less dramatically in cases where not all parties have been able to participate actively in an informal group, or where an overlooked problem is found in the text presented for adoption. It is not uncommon for a party (even one that has participated actively in the informal group) to identify a concern with a draft decision circulated for adoption in plenary, even if the text has previously been debated at length and agreed in an informal group. If this occurs, the Chair will often try to find a fix to the language through brief debate in plenary, rather than reconvening the informal group. At SBSTA 11 in 1999, for example, China objected to a phrase in the draft guidelines for preparing Annex I party national communications that were being presented for final approval by the SBSTA plenary, even though these had been exhaustively negotiated in an informal group. The SBSTA Chair sought to find an acceptable textual amendment in the plenary meeting, surmising that reconvening the informal group would take too long, would convey an importance to the issue that was not warranted, and might incite other parties to raise additional issues with the text. Keeping the debate in an open plenary helped maintain pressure on all parties to agree to a solution, which the SBSTA Chair was eventually able to broker.

It is important to underline, once again, the distinction between plenary meetings of the COP and subsidiary bodies. COP plenary meetings are highly formal, ceremonial events, which are almost always confined to prepared statements – often indeed to stake out positions, 'let off steam' or 'play to the gallery' – and the taking of decisions, including on the organization of work. Otherwise, COP plenary meetings are tightly choreographed by the secretariat to avoid unexpected events and spontaneous debate (Yamin and Depledge, 2004). This does not mean that COP plenary meetings always unfold sedately and without excitement. The contentious nature of the climate change negotiations ensures that moments of high drama have, on occasion, punctuated the solemnity of COP proceedings. A good example was the presentation of the Geneva Ministerial Declaration to the COP plenary at COP 2 in 1996, which provoked heated interventions, followed by thunderous applause when the COP President proposed taking note of the text. Another example was the impassioned debate that burst forth during a COP 3 plenary meeting following the tabling of a proposal by New Zealand on the future development of the regime, including a timetable for the negotiation of developing country commitments. Most delegations were unaware that this proposal would be tabled, and the organizers' plan to avoid debate by moving onto other issues was thwarted by the large number of requests for the floor and the strong interventions of parties.

COP working bodies

Notwithstanding these heated exchanges, the point remains that the plenary functions of taking initial interventions on substantive agenda items, stock-taking, or serving as bargaining forums of last resort, are usually referred by the COP to one or more 'working bodies'. The most common approach has been for the subsidiary bodies to serve as working bodies of the COP, when these have been meeting in parallel with it. However, for the finales of major negotiating rounds, the organizers of the negotiation process have often opted to convene an alternative working body (Yamin and Depledge, 2004). Table 9.2 summarizes the different approaches used to date, which are discussed further below.

Table 9.2 *Working bodies of the COP*

COP	Working body
1	Committee of the Whole
2	Subsidiary bodies
3	Committee of the Whole
4 and 5	Subsidiary bodies
6	Week 1: Subsidiary bodies
	Week 2: Informal high-level plenary
6 (part II)[1]	(Informal negotiating groups)
	Last days: The Group (limited membership friends group).
7[2]	(Informal negotiating groups)
	Last days: The Fez 1 Group (limited membership friends group).
8, 9 and 10	Subsidiary bodies

Note: Shading indicates the finale of a major negotiating round.
Note: 1 The subsidiary bodies did meet in parallel with COP 6 (part II), but did not take up any issues under the post-Kyoto negotiating agenda.
Note: 2 The COP plenary did refer some issues to the subsidiary bodies, but these were not directly related to the post-Kyoto negotiations.

Committee of the Whole

At COP 1 and COP 3, the COP convened a so-called *Committee of the Whole* (CoW) essentially consisting of a hybrid between a formal and an informal group. In the case of COP 3, a CoW was convened under Chair Estrada (rather than the COP President) to finalize negotiations on the Kyoto Protocol on behalf of the COP. Like an informal group, it met in a much smaller room than the cavernous main plenary room, often long into the night, sometimes without interpretation and without written record in the COP report. Like a formal plenary, however, it met in the presence of NGOs and the media, using the traditional seating arrangements and formal rules for the conduct of business, and with sound recordings that are still kept at secretariat headquarters. At the plenary meetings of the CoW, which were held every day, the Chairs of the various informal groups were asked to report on their work, and delegates were given the opportunity to comment. The CoW plenary thus served as a critically important forum for stock-taking,[3] and eventually became a highly effective forum for last resort bargaining. Most of the final text of the Kyoto Protocol was subject to final bargaining and deal-making in a CoW plenary, including the articles on the CDM and general commitments for all parties,[4] which had been laboriously negotiated in informal groups.

Informal high-level plenary

The organizers of the negotiation process used a very different approach for COP 6. The session was divided clearly into two, with the *subsidiary bodies* acting as working bodies of the COP in the first week, and a so-called *informal high-level plenary* (IHLP) convened by the President in the second week. This separation of the COP into two segments had important implications for the time management of the session, as discussed in Chapter 12.

The IHLP convened in the second week retained features of both formal and informal groups, but a slightly different mix of these to that of the CoW discussed above. Like a COP plenary, the IHLP was chaired by the COP President and with interpretation and sound recordings. However, traditional seating arrangements were abandoned. Instead, parties were asked to sit with their negotiating coalitions and were only granted two seats (a great restriction for all but the smallest delegations). NGOs and IGOs were allowed to attend, but there was only limited seating for them. Moreover, the COP President made it clear that he intended the informal *high-level* plenary to be just that: a forum for debate among *ministers*, or at least heads of delegation. This forum did not work well for a variety of reasons. Firstly, there was confusion over its nature. The term 'informal high-level plenary' was an unknown and uncertain quantity, and some parties were genuinely confused as to what they were expected to do there. The ministerial emphasis did not help, given the differing expectations of delegations on the role of their ministers (see Chapter 13). Furthermore, the unorthodox seating arrangements, although they did not raise objections, were also politically ill-advised. Because the negotiating coalitions were forced to sit together, this made it more difficult for political alliances to be forged, broken and reworked in a fluid manner. Some members of the G-77, for example, might have preferred to sit apart from the group on certain issues, but could not do so in a politically subtle manner, because of the defined seating space.

Friends group

At COP 6 (part II) and COP 7, the Presidents initially dispensed with an intermediate working body, directly convening a small number of informal negotiating groups, each charged with a cluster of issues (Yamin and Depledge, 2004). However, in the last days of both these sessions,[5] a limited membership 'friends' group was convened by the Presidents, to oversee the final negotiations, bring together the various strands into a single, coherent document, and (at COP 7[6]) approve the final agreement before presentation to the COP plenary. The outcome of this approach was mixed, in line with the shortcomings (and indeed benefits) of friends groups, which are discussed further below.

Informal arenas: informal groups and informal consultations

The bulk of negotiations in the climate change regime, and indeed in other intergovernmental regimes, takes place in a variety of informal arenas, which are convened by plenary meetings – of the COP, subsidiary bodies or other working body – to address specific agenda items and report back to plenary (Yamin and Depledge, 2004). Their main advantage is that, because they are not bound by rules for the conduct of business, they can serve as more conducive and efficient settings for negotiation. In line with established practice in intergovernmental negotiations, delegates can explore options in informal arenas without being bound by them, but once a deal is struck in the informal arena, it should not be reneged upon when the deal is taken back to the formal plenary for adoption (see Iklé, 1964; Pruitt, 1981). Parties are thus able to talk more freely in informal arenas, knowing that the risks are lower as a proposal will not necessarily be taken as a commitment and no final decisions will be taken (Yamin and Depledge, 2004). As Benedick, a veteran of the ozone negotiations, puts it, 'the aura of informality encourages posing hypothetical questions and advancing unorthodox answers' (Benedick, 1993, p238). Lang concurs that 'informal proceedings have a much stronger impact on the final outcome of negotiations than more or less public debates' (Lang, 1989b, p40).

Inevitably, however, informal arenas face an inherent tension between maximizing efficiency on the one hand, and forsaking the procedural equity and transparency safeguards of formal arenas on the other. The key to the acceptability and effectiveness of informal arenas therefore lies in their careful management. As one interviewee commented:

> if you manage it [the convening of informal groups] in a very bad way, you get complaints. But if you manage it in a very smooth...or discreet way, you will find that they [delegates] accept this. So...the whole thing is the management.

The absence of set rules for the conduct of business in informal arenas opens up considerable room for improvisation in their design in order to respond to the needs of the negotiation process. A variety of informal arenas have therefore been used over time, differing in their size, mandate, negotiating procedures and degree of formality. Among this variety, we can identify two broad categories: informal groups and informal consultations. Each of these is discussed below.

Informal groups

Informal groups are known by a variety of terms (Yamin and Depledge, 2004). '*Contact group*' is the most common in the climate change regime, and is habitually used to refer to informal groups convened by the subsidiary bodies. During major negotiating rounds, different terms have been used. During the Kyoto Protocol negotiations, for example, Chair Estrada chose to use the term '*non-group*' for the informal groups he established in 1997. This term, unprecedented in the climate change regime, sought to underscore that the groups had no formal status, thereby responding to concerns of parties that bargaining and deal-making should not yet commence (see Chapter 12). At COP 3, however, the name given to these groups was changed to '*negotiating groups*' in order to emphasize that bargaining and deal-making would now begin in earnest, and sub-groups convened under the four main negotiating groups were termed 'contact groups'. 'Negotiating group' was also used to refer to the handful of informal groups convened for the finale of the post-Kyoto negotiations at COP 6 (part II) and COP 7, again with spin-off groups known as 'contact groups'. At COP 6, the term '*cluster groups*' was employed, this time underlining the fact that two of the four groups were dealing with a cluster of related issues.[7] The term '*drafting group*' is also sometimes used in the climate change regime, where the informal group has a specific mandate to draft a particular piece of text. '*Working groups*' were convened during negotiations on the Convention itself, but the term is no longer in common usage in the climate change regime, as it is considered to denote a level of formality on a par with a Committee of the Whole that is usually not desired for informal groups. '*Joint contact groups*' of both subsidiary bodies have also been held on issues under the joint responsibility of both the SBSTA and the SBI. This includes negotiations on the flexibility mechanisms and adverse effects during the post-Kyoto negotiations, and occasionally also at other times on such issues as national communications, technology transfer and activities implemented jointly (AIJ).

Although no set rules exist to govern informal groups, established practice has evolved in the climate change regime such that, whatever their name, informal groups tend to share certain key characteristics (see also Yamin and Depledge, 2004). Most importantly, informal groups in the climate change regime are *open-ended*, that is, open to participation by all parties. They are usually chaired by a delegate other than the Chair of the convening body, with a practice having emerged since Kyoto of appointing Co-Chairs, one each from an Annex I and a non-Annex I party. Negotiations are conducted in English-only (interpretation has only been provided in exceptional cases) and working documents are rarely translated. NGOs are now allowed to attend open-ended contact group meetings as observers (unless at least one third of parties object), but the group Chair may request them to leave at any time, usually when negotiations enter a delicate stage (see below). Informal groups are not bound by traditional UN meetings times, and will often meet into the night. However, established practice requires that no more than two meetings (including plenary meetings and informal groups) be held at any one time (see Chapter 7). Meetings of informal groups are advertised in the official daily programme of meetings and on electronic noticeboards.

Informal consultations

An alternative type of informal arena consists of 'informal consultations'.[8] Although the distinction is not always clear, informal consultations are convened by the *presiding officer* (COP President or SBSTA/SBI Chair), in contrast with informal groups, which are established by the COP or subsidiary bodies. The presiding officer typically invites a delegate (sometimes two) to consult on a particular topic and report back to plenary, but advertised meetings are not held. The expectation, instead, is that the consulting Chair will discuss the issue at hand in private with representatives of the main negotiating coalitions and interested delegations in order to forge a consensus. An informal consultation is therefore a much more private process than a contact group or any other type of informal *group* (Yamin and Depledge, 2004). If meetings are held, NGOs will not be invited, and venues and times will not be advertised. Importantly, informal consultations are not bound by the no-more-than-two-meetings practice.

In general terms, informal consultations, rather than informal groups, are convened for two contrasting reasons (Yamin and Depledge, 2004). Firstly, on more technical or procedural issues, or on those where there is limited general interest, so that a full-scale informal group is not warranted. Informal consultations, for example, were convened at COP 3 to find an alternative terminology to 'emission budgets' and, among many other examples, at SBSTA 19 on inventory issues and registry systems. Alternatively, informal consultations are sometimes convened for diametrically opposite reasons, that is, on sensitive political issues, where it is feared that a more open group negotiation might given dangerous prominence to the issue, or trigger unproductive confrontation. Some of the many examples that could be cited include consultations on voluntary commitments for developing countries at COP 3, on the relationship between the climate change and ozone regimes at SBSTA 17, and on the Canadian proposal on cleaner energy and the implementation of Kyoto Protocol Article 2.3 (adverse effects), two controversial and politically (if not substantively) linked items, at SBSTA 19. Informal consultations are similarly used where a more discreet process is appropriate, notably for elections (e.g. to the COP Bureau or CDM Executive Board) or on issues relating to a specific country, such as Turkey's request to be deleted from Annex I and Croatia's special circumstances as an EIT.[9]

Informal consultations have been used extensively in the 'finales' of the climate change regime's main negotiating rounds, especially post-Kyoto. In the final stages of the political segment of COP 6 (part II), for example, four sets of informal consultations were convened under the main friends group, each 'facilitated'[10] by a delegate (in three cases, by ministers) appointed by the President. Convening informal consultations in this way (rather than an informal group) provided more flexibility to the appointed facilitators, as they were under no obligation to convene an open meeting and could conduct the process as they saw fit without any procedural constraints.

The functions of informal arenas

Informal arenas – both informal groups and informal consultations – are used a great deal in the climate change regime. The number convened at each session varies, depending on the agenda and burden of work. Between 12 and 22[11] informal arenas have been convened at each negotiating session since the latter stages of the Kyoto Protocol negotiations in 1997. Despite widespread perceptions among interviewees to the contrary, there is no clear trend towards a rise in the number of informal arenas. Just as many informal groups and consultations, 16, met during the July sessional period in 1997 as did during SBSTA/SBI 20 in 2004. It is arguable, however, that informal arenas are being used more intensively. In the above example, the same number of informal groups are now being convened by just two subsidiary bodies, rather than the SBSTA, SBI, AGBM and AG13 in 1997. In addition, almost all informal groups are now convened to develop conclusions and decisions on entire agenda items, not just to discuss a particular tricky sub-issue, and are meeting sooner in the session with increasingly minimal plenary discussion.

Promoting a bargaining atmosphere

The main purpose of informal arenas is to promote a more constructive bargaining atmosphere. An important element in this is simply the smaller number of delegates taking part in informal arenas compared with plenary meetings. The number of people actually present at an informal group will vary widely depending on the interest in the topic. The flexibility mechanisms contact group, for example, which met during the post-Kyoto negotiations, was always very well attended, with 100 or more delegates, whereas less high-profile groups, such as those on methodological issues, tend to attract 20 or 30. Informal consultations will often involve just a dozen or so parties, representing the main negotiating coalitions and other major players on the issue.

Negotiations in informal groups are held in smaller, less overpowering settings than the main plenary meeting rooms, without prescribed seating and often without the use of country flags (Yamin and Depledge, 2004). This helps to create a more intimate atmosphere that can encourage constructive debate. Informal groups, especially those that are repeatedly convened session after session, will often develop a liking for a particular setting, which they feel generates a productive dynamic. In smaller rooms, seating may be organized in a square arrangement, rather than in rows, enabling face-to-face conversation without microphones, which can boost spontaneity and understanding. Allowing delegates to choose their own seating also encourages more active bargaining, as parties can sit close to allies, thus enabling them to confer immediately and exchange views, thereby speeding up the negotiation process. Interestingly, not all interviewees agreed on the benefits of face-to-face negotiations. For some, this can provoke greater confrontation and face-saving behaviour, compared with seating in rows, where inability to see the face of the negotiating opponent can actually help parties to back down from their positions and accept compromises.

Specialization

Establishing an informal group or informal consultation on a particular issue introduces a degree of specialization to the negotiations, and only those parties and delegates specifically interested in the topic will participate. Parties not so interested will often be happy to be represented in the informal arena by the representative of their negotiating coalition. This specialization allows delegates to focus their efforts on a single issue, unencumbered by the complexity of the rest of the negotiations, while building up a community of colleagues with particular expertise, who will get used to working together, sometimes across several sessions.

Delegation of chairing

Delegating the task of chairing helps to spread ownership for the negotiations among a wider group of individuals. A greater number of delegates from a wider range of countries feel a personal and professional responsibility for bringing the negotiations (at least those they were working on) to a successful conclusion. It also makes for a more inclusive process, as the subsidiary body Chair or COP President is not the only one arbitrating between the opposing views of parties. It also frees the subsidiary body Chairs and COP President to hold their own private talks, and think strategically about the negotiations. Sometimes, however, the subsidiary body Chair may choose to chair a contact group or informal consultation, if the issue under discussion is particularly sensitive or important. Chair Estrada, for example, chose to chair the negotiating group on quantified emission limitation and reduction objectives (QELROs) himself at COP 3, in view of its centrality to the negotiation process. Similarly, SBSTA Chair Thorgeirsson decided to chair the contact group on the IPCC's TAR at SBSTA 19, given its crucial importance in setting the agenda for the future work of the SBSTA.

Being invited to chair an informal group or to conduct informal consultations is considered an honour, and the Chair will therefore exert every effort to reach agreement, and not have to report failure to plenary. Choosing whom to invite to chair an informal group or conduct consultations is a delicate task, and must take account of regional balance in the nationalities of Chairs and other political considerations, as well as chairing ability. Issues relating to the appointment of Chairs are discussed in more detail in Chapter 4.

English-only negotiations

The conduct of negotiations in informal arenas in English only inevitably places non-Anglophone nations at a disadvantage, especially less well-resourced delegations that do not comprise individuals with language training. The greater spontaneity and quicker pace of informal negotiations redoubles this disadvantage. As Kaufmann explains, 'while the language advantage or disadvantage is not so serious for statements prepared at leisure...it can become acute when an impromptu intervention suddenly becomes necessary, and a delegate may be groping for the right words' (Kaufmann, 1989, p175).

Informal groups have sometimes been provided with interpretation on an exceptional basis. A case of this occurred at COP 6 (part II), where, at the request of a small number of developing countries, a limited number of interpretation slots was provided for the negotiating groups. These slots were taken up especially by the negotiating group on finance, an issue of particular concern to developing countries, illustrating the fact that developing countries are especially affected by the absence of interpretation. Interestingly, however, some negotiating group Chairs, notably the negotiating group on compliance, declined to use the interpretation offered to them, claiming that it would actually slow down negotiations.

The question of language is a perpetual problem in intergovernmental forums. It is impossible to provide interpretation for every single meeting and, even if such interpretation were provided, the fact is that negotiations would almost certainly still take place in a single language. In a context where finding the precise word is paramount, highly technical and novel terms are being discussed, and spontaneous interactions are necessary, negotiation in more than one language is simply impractical. The crux of the problem thus lies more in the pervasive global imbalance of wealth and resources than in a fundamental flaw in the organization of negotiations. Larger or better resourced non-Anglophone delegations with training in English, more time to devote to the negotiation process and the capacity to translate documents into their own language are not as handicapped as the poorer countries with smaller delegations, such as Francophone Africans. A Chinese interviewee explained: 'for a big delegation like China, it [the lack of interpretation] was absolutely no problem...everybody works in English now...those technical terms, they are hard to translate, you can't use the Chinese version'.

Monolingual negotiations pose problems even for some large delegations, however, including that of the Russian Federation. The linguistic isolation felt by the Russian Federation is believed to have stoked the sense of neglect that contributed to its insistence on a generous emission target in the Kyoto Protocol, and later to its demands to renegotiate its allocated sink credits under the Bonn Agreements, and even, perhaps, to its prevarication on ratifying the Kyoto Protocol (see also Chapters 7 and 11). Moreover, for many countries, the provision of interpretation and opposition to English-only negotiations is a matter of principle, which in turn raises opportunities for procedural obstruction. Managing such language issues is thus an important task for the organizers of the negotiation process, despite the implicit acceptance of the inevitability of English monolingualism in the final stages of a negotiating round.

Scheduling

Holding parallel meetings of informal groups intensifies the use of time, enabling negotiations to advance simultaneously on different fronts. The established practice, however, of holding no more than two meetings at any one time, including plenary and informal group meetings, serves as a safeguard to uphold procedural equity and transparency for small delegations. Although this practice was prevalent before Kyoto, it appears that it was only applied to the four bodies in operation at that time – SBSTA, SBI, AGBM and AG13 – *separately*, rather than in combination. The daily programme for the two-week sessional period in July 1997, for example, reveals many instances of three, sometimes four, informal groups of the various bodies meeting at any one time. Delegates, however, especially from developing countries, have sought the application of this rule more zealously in the post-Kyoto era and, crucially, *across both subsidiary bodies together*. The need to fit more and more informal groups into just two parallel slots has gradually expanded meeting hours later into the night, and also squeezed the traditional three-hour meeting slot into just two or even one and a half hours (Yamin and Depledge, 2004). The secretariat drafts the schedule for informal group meetings, and in doing so, must ensure that all groups receive equitable treatment. The subsidiary body Chairs or, at COP sessions, the COP Bureau, will approve the schedule of informal group meetings.

The fact that informal consultations bypass the no-more-than-two-meetings practice that covers informal groups further intensifies the use of time. This is crucial to managing the complexity of the climate change negotiations. It would be logistically impossible to resolve the full spectrum of issues on the agendas of the climate change bodies with only two groups meeting in parallel. Where the schedule has been particularly tight, informal consultations have sometimes been convened rather than informal groups, in order to avoid the constraints imposed by the no-more-than-two-meetings practice.

Participation of NGOs
A feature of informal arenas that has changed over time is the treatment of NGOs. Prior to COP 4, there were no rules specifying whether or not NGOs were allowed to observe, or indeed speak at, informal group meetings, leaving this entirely to the discretion of the Chair of the convening subsidiary body or the Chair of the informal group. This led to considerable variation in the practices of informal groups. The SBSTA, for example, adopted a more open practice, with NGOs usually permitted to observe contact groups. Chair Estrada, however, took a stricter line in the AGBM, barring NGOs from the non-groups convened in 1997. The rather ad hoc and unpredictable treatment of NGOs was remedied at COP 4 by decision 18/CP.4 on attendance of intergovernmental and non-governmental organizations at contact groups. This decision establishes that the convening body should invite IGOs and NGOs 'to attend as observers any open-ended contact group', unless at least one third of the parties object at the time of its establishment. Chairs of contact groups, however, retain the right to close the group to IGOs and NGOs at any time. Participation by NGOs is discussed in more detail in Chapter 14.

Obstacles for smaller delegations
Although the convening of multiple informal groups and consultations has advantages for the efficiency of the process, it inevitably has a detrimental

effect on procedural equity and transparency. Small, especially non-Anglophone, delegations – mostly from developing countries and EITs – often lack the resources to send representatives to the groups of greatest interest to them, or to keep track of the many threads of the negotiation. Moreover, while larger delegations can appoint specialists to different sub-groups, thus taking advantage of the broad specialization noted above, this is usually impossible for small delegations.

Taking the example of the Kyoto Protocol negotiations, one interviewee recalled, 'many delegates were not involved in small groups, they were left out, they were left not knowing what was going on'. Another interviewee noted, 'many smaller delegations could not attend all the parallel informal group meetings ... so they say, "Well, we didn't participate fully in this group", so whatever outputs you get from that group is viewed with some suspicion'. For another interviewee, the convening of smaller groups contributed to the domination of the larger, more powerful delegations over the whole Kyoto Protocol negotiation process:

> As the issues developed, there were smaller negotiating groups ... and as the groups got smaller ... then we started to lose out on participation and ... it just made it easier for countries who wanted to minimize the outcomes ... I guess the US is the classic example ... they were involved right to the end in the smaller and smaller groups.

Post Kyoto, smaller delegations have become more efficient at easing the problem of multiple informal arenas by working through their negotiating coalitions. The G-77, for example, appoints coordinators for most issues (a practice that started in the latter stages of the Kyoto Protocol negotiations), who negotiate on the basis of a mandate from the Group and report back to it. Such self-help by parties is critical to enabling a broader, if not deeper, participation on a greater number of issues in the negotiations. However, by ceding negotiating power to a group representative, it is inevitable that, for all but the most powerful states, individual national preferences cannot be argued as strongly.

Potential for incoherence and inconsistency

The specialization of the negotiations can result in incoherence and inconsistency in the final outcomes, as even the presiding officers and secretariat find it difficult to keep abreast of developments in the various groups and maintain a big picture view of the process. This situation during the Kyoto Protocol negotiations contributed to some incoherence and inconsistency in the Protocol text, which negotiators have had to contend with in the post-Kyoto negotiations. According to Yamin:

> ... many elements ... took shape in the small hours in 'informal informals' in many different rooms, often meeting simultaneously,

so that few outside the larger delegations could follow the entirety of the negotiations. These factors ... go some way to explaining the ... idiosyncrasies of the text (Yamin, 1998, pp115–116).

This impression was echoed by interviewee responses. One, for example noted:

A lot of negotiations have been done independently in small groups where macro pictures were not able to be conceived ... today many people come and tell you 'Why should you levy [a share of the proceeds of] the CDM [to help vulnerable developing countries adapt to climate change] and not other mechanisms?'... simply because everyone was in smaller groups.

The problem of incoherence and inconsistency was considerably more acute in the post-Kyoto negotiations, given the more disaggregated institutional structure without a single body (such as the AGBM) to oversee the whole process. Informal groups and consultations were used more intensively, and there were strong links between issues. For example, the contact group on LULUCF was devising rules for including LULUCF projects under the scope of the Kyoto Protocol, while the contact group on the flexibility mechanisms had responsibility for negotiating the treatment of LULUCF projects under the CDM. The compliance and flexibility mechanisms groups also had important cross-cutting issues relating to eligibility to participate in the flexibility mechanisms, and redress for any lapses in eligibility. Inconsistency between the informal group outputs that comprised the Bonn Agreements led to considerable problems in the more technical negotiations on detailed texts at COP 7. One important inconsistency concerned the treatment of LULUCF projects in the CDM. Such inconsistencies were formally noted in the COP 6 (part II) report, and the secretariat was forced to issue an inventory of them (see COP 6 (part II) report, part I, paragraphs 47–49).

The secretariat and presiding officers have sought to address the dangers of inconsistency in the negotiation process in various ways. One important means has been through the 'clustering' of issues into single contact groups. A good example here is the contact group on the flexibility mechanisms, which dealt with emissions trading, JI and the CDM throughout the post-Kyoto negotiations. While making for a very heavy agenda for the contact group, this clustering was helpful in ensuring a degree of consistency throughout the rules for the flexibility mechanisms, while reducing transaction costs involved in designing separate provisions for each mechanism.[12] All three flexibility mechanisms share common principles and basic eligibility rules for participation, while the two project based mechanisms – the CDM and JI – also share many similar rules regarding the project cycle and the functioning of their respective bodies (Executive Board and Supervisory Committee). Another means was through the occasional convening of joint meetings of related contact groups. The LULUCF and flexibility mechanisms contact groups, for example, held a joint 'special session' at SBSTA/SBI 13 just prior to COP 6.

A further way in which the organizers of the negotiation process have sought to facilitate coherence is by inviting all the Chairs of informal groups and informal consultations to private coordination meetings with the subsidiary body Chairs (or COP President, if relevant). Chair Estrada, for example, convened regular meetings of the informal group Chairs from AGBM 7 onwards. These meetings proved to be important in addressing problems relating to coherence and integration. For example, at AGBM 8 in late 1997, confusion arose over which non-group had responsibility for the issue of review of implementation, with the result that two non-groups had started to take up the issue in different ways. The coordination meetings were able to clarify which non-group should have responsibility for the issue. This strategy of convening coordination meetings of informal group Chairs continued in the post-Kyoto process where, especially during the finales of COP 6, COP 6 (part II) and COP 7, such meetings were held on a daily, even twice daily, basis.

Another related means is self help by the secretariat and parties. The secretariat had an active coordination team in place during the post-Kyoto negotiations to try to ensure coherence and consistency between negotiations on the various related issues, alerting presiding officers and senior secretariat staff to potential problems. Similarly, all negotiating coalitions and most large delegations were usually represented in the informal arenas on all issues, albeit by different individuals, and communication within delegations and coalitions could therefore help ensure that coherence problems were spotted.

Such strategies were absolutely crucial in avoiding many pitfalls of inconsistency. The problems that did creep into the texts can be largely attributed to the pace and late hour of the final negotiations, especially at COP 6 (part II).[13]

Friends groups and shuttle diplomacy

Friends groups

In addition to the negotiating arenas explored above that are open to all parties, presiding officers have sometimes invited a limited number of parties to a series of private meetings aimed at advancing the negotiation process (Yamin and Depledge, 2004). Such groups, referred to generically here as 'friends of the President' or 'friends of the Chair', are most often convened by the COP President in the final stages of a negotiating round. They have also occasionally been convened at the initiative of the subsidiary body Chairs. Key friends groups held in the climate change negotiations from 1996–2003 and their main characteristics are summarized in Table 9.3 below.

The convening of such friends groups by the presiding officer is commonplace in multilateral negotiations.[14] Like informal arenas, there are no set rules governing their conduct, opening up considerable room for improvisation on the part of the presiding officer as to what kind of group s/he wishes to convene (if at all) and how to use it (Yamin and Depledge, 2004). Friends groups may be convened inter-sessionally to prepare for the upcoming session as well as during negotiating sessions, while the convening presiding officer may choose to delegate the chairing of the group to a colleague or colleagues (e.g. see cases at COP 2 and COP 7 in Table 9.3). The distinguishing characteristic of all friends groups, however, is that they are *limited membership bodies*; that is, only invited delegates are permitted to attend, to the extent that security guards may be placed at the door to deter uninvited individuals. The size of the invitation list has varied considerably from group to group, from attendance by around 15 individuals (the AGBM Expanded Bureau) to over 60 (friends at COP 6, COP 6 (part II) and COP 7). The point, however, is that friends groups seek explicitly to exclude the majority of delegates, in order to increase the efficiency of decision-making in the final stages of difficult and sensitive political negotiations. As one interviewee put it, 'big decisions mean small groups'. The major exception to this was the open-ended informal high-level consultations convened in the Netherlands in the run up to COP 6 (part II), where participation was open to all parties. This was an extremely innovative gathering, as discussed in Box 9.2 below.

In the interests of promoting efficiency, friends groups are subject to even fewer procedural equity and transparency safeguards than the informal arenas discussed above. Negotiations always take place in English only. It is indeed partly for this reason that the Arabic and French speaking COP 7 President El Yazghi chose to appoint English-speaking ministerial 'co-facilitators' to lead the COP 7 friends group on his behalf. The conduct of business is entirely at the discretion of the Chair, and meetings may be held anywhere, at anytime, and at a moment's notice. NGOs are not permitted to attend, and meetings are rarely openly advertised (Yamin and Depledge, 2004).

Friends groups have been known by many different names in the climate change process (see Table 9.3 below), reflecting the uniqueness of each group and its composition, along with the controversy that inevitably surrounds meetings to which not all parties are invited. Indeed, the term friends itself has usually been rejected, because the implication could be that other parties are not the friends of the Chair or President. Names implying that the group could have any kind of formal standing must also be avoided. Most commonly, friends groups have been known by the name 'Expanded Bureau' (Yamin and Depledge, 2004), where the composition of the group has been founded on the COP Bureau membership, drawing on the legitimacy of the Bureau as a formally elected group (on composition, see below). Alternatively, the name of the meeting place for the group has been used, such as the 'Muiden group' prior to and at COP 6, or the 'Fez 1 group' at COP 7. Even the AGBM Expanded Bureau was known for a short while as the 'Trattoria group', as it first met for dinner in an Italian restaurant. The case of COP 6 (part II) is interesting. Here, given lingering unhappiness over the failed friends group at COP 6, any name was deemed inflammatory, so that the group was simply called 'The Group'.

The fundamental aim behind the convening of friends groups is to increase the efficiency of the negotiation process. The reasoning is that, because fewer people are present, the process will be simpler and more efficient, and it will be easier to reach an agreement. Linked to this is the assumption that delegates will talk more freely and frankly in a small, closed, intimate setting than they otherwise would in a more open forum. The price to pay for such apparent improvements in efficiency is, of course, an erosion of transparency and procedural equity, as parties cannot participate as actively in the negotiation process, observers are excluded, and no official records are kept of proceedings.

The presiding officers and secretariat often go to considerable lengths to create an inspirational setting and sense of occasion that will be conducive to constructive talks. At COP 6, for example, the Dutch Presidency convened the second meeting of the 'Muiden Group' in the historic Riversaal, a room imbued with political significance for the Dutch people. At COP 3, steps were also taken to instil a sense of occasion, responsibility and urgency among participants in the 'Ministerial Group'. After the first meeting where debates had been disturbed by delegates moving in and out of the room, Chair Estrada requested the doors to be kept shut with UN security outside. Participants were discouraged from leaving the room and no one was allowed in.

Composition

The composition of friends groups in the climate change regime and the method for selecting participants has evolved over time. During the Kyoto Protocol negotiations, the convening President or Chair simply drew up a list of invited parties, which, of course, had to be carefully managed so as to minimize controversy. One means of doing this, used by Chair Estrada during the AGBM process, was to base the invitation list on the membership of the COP (and also the AGBM) Bureaux – hence the name, Expanded Bureau. The remaining invitees – the 'expansion' – were representatives of key negotiating coalitions not covered by the Bureau structure (e.g. G-77 Chair, EU President), and key players identified by the organizers by virtue of their strong positions and leverage power (e.g. China, India, US), or the personal influence of individual delegates (e.g. Antigua and Barbuda). Although the Expanded Bureau was widely accepted in the AGBM, there were still protests from some parties excluded from the meetings. The practice of using an 'expanded Bureau' – COP Bureau membership plus 'key parties' – has persisted in the climate change regime, and has been used to structure most friends groups convened for consultation purposes (on functions, see below).

In the post-Kyoto period, it has become less acceptable for a convening Chair or President to draw up an invitation list of parties on a 'top down' basis, especially for friends groups used as bargaining arenas (see below). Instead, the trend has veered towards a more 'bottom up' process, that is, with invitations extended to the negotiating coalitions, and the coalitions themselves selecting their participants. It is necessary, however, to set a quota of seats for each negotiating coalition; the alternative is inevitably a very over-crowded room and an inefficient process, as happened at the COP 6 friends group. At COP 6 (part II), agreement was reached on an allocation of seats for each negotiating coalition, which was subsequently applied without question to structure the 'Fez 1 Group' at COP 7 (see COP 6 (part II) report, part I, paragraph 33).

Establishing a quota of seats, then allowing coalitions to identify their own representatives, has the advantage of being a more democratic and open process, disposing of the responsibility for deciding on representation to the

Box 9.2 Making up for the failure of COP 6: The open-ended informal high-level consultations in Scheveningen, the Netherlands, June 2001

The friends group that met for two days between COP 6 and COP 6 (part II) in Scheveningen, the Netherlands was, and remains, unprecedented in the climate change regime. An open invitation was issued to all parties, with funding provided to eligible parties in the same way as for the formal negotiating sessions, thanks to financial support from the Dutch government. Some 350 delegates representing 130 parties attended, including many ministers. Observers, however, were not admitted.

The aim of the meeting was two-fold. Firstly, COP 6 President Pronk wanted to present his 'consolidated negotiating text', released a fortnight before, and elicit the views of parties on whether it could provide a basis for reaching agreement at COP 6 (part II). Secondly, President Pronk wished to consult on the organization of work at COP 6 (part II), to avoid the tortuous debate on process that had so crippled COP 6. In this sense, the Scheveningen meeting appeared to represent an attempt to atone for the perceived mistakes of COP 6. Having been accused of releasing a complex text too late at COP 6 (see Chapter 11), President Pronk now sought to consult on his new text before the negotiating session. In response to charges of errors of judgement and hesitation in his organizational decisions at COP 6, President Pronk now sought to consult on these prior to COP 6 (part II). Most importantly, President Pronk wanted the Scheveningen meeting to be open to all parties, so he could not be accused of lack of transparency and could justify the decisions that he later took.

The meeting was certainly useful, not just in advancing its explicit aims discussed above, but also in providing an opportunity for delegates, most of whom had last seen each other on the disastrous last night of COP 6, to meet up, exchange views and rebuild relationships. It was also the first opportunity for a wide group of parties to meet following the repudiation of the Kyoto Protocol by the US. Importantly, the schedule included meetings of the various negotiating coalitions, along with a preparatory meeting of the G-77 on the eve of the consultation. Although the Scheveningen consultations did not lead to, or provide any indication of, constructive changes in position, the fact that parties began serious work immediately on the first day of COP 6 (part II) is due in no small part to the confidence building that emerged from that meeting.

coalition Chairs (Yamin and Depledge, 2004). It is, of course, easier for some coalitions than others to choose their representatives. It is straightforward, for example, for the EU, given its very cohesive position and formal rotation of the Presidency. The G-77, however, must cover a very large spectrum of views; the sight of the Iranian delegate – Iran held the G-77 Chair – being mobbed by G-77 members trying to obtain one of the 19 badges that he was distributing for access to the friends group was one of the enduring scenes of COP 6 (part II).[15]

Table 9.3 *Key friends groups convened under the COP or AGBM in the climate change regime up to the adoption of the Marrakesh accords*

Session	Name	Function	Composition	Chair
COP 2	Friends of the President	Bargaining on declaration	Invitation list of key Parties	Minister Marchi (Canada)
AGBM 6-8	Expanded Bureau	Consult on process, some substance	Bureau + invited Parties	AGBM Chair Estrada
COP 3	Group of 10 (throughout) Informal consultation (only one meeting)	Share information Bargaining	Bureau + invited Parties Invited Parties	CoW Chair Estrada COP 3 President Ohki
	Ministerial Group (last 3 days)	Bargaining	Bureau + invited Parties	CoW Chair Estrada
COP 4	Friends of the President (last 2 days)	Bargaining	Bureau + invited Parties	COP 4 President Alsogaray
COP 5 – COP 6 inter-sessional	Informal high-level consultations (April 2000, New York)	Consult on preparations for COP 6	Bureau + invited Parties	COP 5 President Syzsko
	Informal high-level consultations (June 2000, Warsaw)	Consult on preparations for COP 6	Bureau + invited Parties	COP 5 President Syzsko
	Informal high-level consultations Muiden group (Oct 2000, Muiden, Netherlands)	Consult on substance, based on non-paper by COP 6 President	Bureau + invited Parties	COP 5 President Syzsko/ COP 6 President Pronk
COP 6	Muiden group (middle Sunday)	Consult on substance	Bureau + invited Parties	COP 6 President Pronk
	Frans Hals group (last day)	Bargaining	Loose invitation to coalitions	COP 6 President Pronk
COP 6 – COP 6 (part II) inter-sessional	Informal high-level consultations (April 2001, New York)	Consult on non-paper 'new proposals by the President'	Bureau + invited Parties	COP 6 President Pronk
	Open-ended Informal high-level consultations (see Box 9.2) (June 2001, Scheveningen)	Consult on: 'Consolidated negotiating text'; process at COP 6 (part II)	Open-ended. 350 delegates from 130 Parties attended	COP 6 President Pronk
COP 6 (part II)	Extended Bureau (first days)	Consult on process	Disputed quotas for coalitions	COP 6 President Pronk
	The group (last 3 days of political talks)	Bargaining	Agreed quotas for coalitions	COP 6 President Pronk
	Compliance group (last few hours)	Bargaining	Loose invitation to coalitions, very small room	COP 6 President Pronk
COP 7	Fez 1 group (last 3 days)	Bargaining	Agreed quotas for coalitions	Ministers Moosa (South Africa), Roch (Switzerland)

Structuring participation in a friends group around negotiating coalitions can, however, have the undesired effect of actually hardening and perpetuating coalition positions (Yamin and Depledge, 2004). It inherently introduces a more adversarial mode of debate by explicitly pitting coalitions against each other, while constraining more fluid interaction and alliance building between members of various coalitions. The LDCs, for example, or AOSIS, could not speak on their own behalf at the negotiating table without the danger of appearing to diverge from the G-77 position, something that would be very politically difficult for them to do.

Concerns at exclusion can be addressed to some extent by providing seats for other parties to observe proceedings, enabling them to speak directly with their representative at the table if necessary. Such provisions were made at COP 6 (part II) and COP 7, where the medium-sized room accommodated 50 or so observers. It also proved helpful to allow rotation between representatives at the negotiating table at any time. For large groups like the G-77, this is important in enabling input by experts on their specific issue. As with informal arenas discussed above, an important means of countering the sense of exclusion felt by the bulk of parties not present at the friends group, along with NGOs, can be to make regular reports to a plenary meeting.

Functions of friends groups

The various friends groups convened in the climate change regime have had two main functions: consultation or bargaining (Yamin and Depledge, 2004).

Consultation forums

Friends groups established as *consultation forums* essentially serve as 'focus groups'; that is, the President (or Chair) has access to representatives of the main negotiating coalitions, and can thus canvass their views or advice on procedural and substantive issues. Because meetings are held in private and behind closed doors, the expectation is that participants will be candid and forthcoming in the advice given to the President, faithfully relaying the views of their negotiating coalition and putting forward constructive suggestions. This function, of course, is similar to that of the official COP Bureau (see Chapter 5). Friends groups, however, are larger and more representative of the substantive negotiating coalitions, and are therefore more effective as consultation forums than the COP Bureau itself.

Convening a friends group as a consultation forum prior to each COP session has now become almost an established practice in the climate change regime. This dates back to COP 5, where the COP President received an explicit request through decision 1/CP.5 'to take all necessary steps to intensify the negotiating process' in preparation for COP 6, with the assistance of the COP Bureau (see also Chapter 12).[16] Two Expanded Bureau meetings were thus hosted by the Polish COP 5 Presidency, followed by the 'Muiden Group' meeting hosted by the Dutch President designate of COP 6.

Friends groups convened as consultation forums tend to have two main purposes. Firstly, they can serve as useful forums for the President to consult on *process and procedural matters*, that is, to seek advice on plans for the organization of negotiations. During the Kyoto Protocol negotiations, Chair Estrada regularly used meetings of the AGBM Expanded Bureau to test out his intentions for the negotiation process, for example, whether parties would be prepared to give him a mandate to prepare a Chair's text. Similarly, at COP 6 (part II), President Pronk convened an Expanded Bureau specifically to advise on the negotiation process. Although it met only twice, this body was important in sanctioning the document on which ministers would negotiate, as well as endorsing the launch of another friends group, this time with a bargaining mandate.

Secondly, friends groups can be used to exchange views on *substantive options* in the negotiation process, albeit with a clear understanding that no negotiation will take place. This, for example, was the intent behind the 'Muiden group' of ministers that met twice, immediately prior to and then during COP 6. At both meetings, President Pronk presented a 'non-paper' setting out issues and options in the negotiation process (see Table 11.4). He did so to provoke debate aimed at generating a better mutual understanding among parties of one another's concerns, along with hints at the viability of different potential solutions. Similarly, the AGBM Expanded Bureau meetings, especially those held inter-sessionally, were used to provoke more uninhibited substantive debate, with parties taking advantage of the opportunity to explain their preferences and the rationale behind these more fully, even if they did not move from their positions.

Frank exchanges in friends groups are also useful to the organizers of the negotiation process. A secretariat interviewee noted:

> Hearing what people said helped us to find the compromise language to bring parties to agreement. The opportunity to hear delegations in the informals meant that we could do this better. It gave us advice to craft a text. We don't have such interchange in big meetings.

Bargaining forums

Friends groups used as *bargaining forums* are inherently more controversial, as few parties are happy to relinquish their right to participate fully in actual bargaining. Where friends groups have been convened for bargaining purposes, this has always been in the very final stages of a major negotiation round, in order to bring the various strands of the negotiations into one forum and strike a final deal.

Friends groups in the climate change regime convened as bargaining forums have enjoyed limited success. The friends group at COP 6 famously catapulted the session into breakdown. Although COP 3, COP 6 (part II) and COP 7 were all successful, their respective bargaining friends groups were largely sidelined as the real bargaining among powerful parties shifted to other, even more informal and smaller settings. At COP 7, for example, the 'Fez 1 Group' developed purely into a forum for the co-facilitators to present proposals and – a few hours later – hear responses from each of the coalitions. The 'Fez 1 Group' was therefore useful, in that it took charge of the final process, brought all remaining strands within one setting, accelerated the pace of negotiations and placed pressure on recalcitrant parties to compromise, but it did not serve as an actual bargaining or deal-making forum. The situation was similar at COP 6 (part II), where 'The Group' proved useful as a more open forum for presenting the President's compromise proposal – the 'Core Elements' paper – and taking collective stock of progress in 'shuttle diplomacy' and unofficial negotiations, but no bargaining actually took place therein. In this case, the President tried to encourage frank exchanges among parties, but these were little more productive than in other arenas, prompting him to convene spin-off informal consultations, and finally, a round of 'shuttle diplomacy'.

Bargaining and successful deal making did take place in the bargaining friends groups at COP 2 and COP 4, but both were later derided for lack of transparency and inclusiveness. The rather secretive and shadowy nature of the friends group that led to the Geneva Ministerial Declaration at COP 2, for example, contributed to the objections that prevented the Declaration's adoption (see COP 2 report, part I, annex, statement by Saudi Arabia). Similarly, at COP 4, Switzerland registered a formal protest 'at the exclusion of many countries from the informal ministerial consultations convened by the President' that reached agreement on the decisions making up the BAPA (see COP 4 report, part I, paragraph 78).

The 'Compliance Group' at COP 6 (part II) was perhaps the most successful friends group in actually striking a deal that was widely accepted as legitimate. This group is unusual among friends groups, however, in that it was convened to negotiate only one issue, and only in the final hours of negotiation; strictly speaking, it was more of an informal consultation than a friends group. Moreover, its work was not universally welcomed; the chief Russian negotiator was momentarily 'conveniently' absent (Lefeber, 2001, p31) when the deal was struck, and subsequently objected to agreement having been reached in English only. This contributed to Russian hostility to the emerging Bonn Agreements, which prevented their swift adoption (see also Chapter 7).

Problems with friends groups

There are several reasons for the relative ineffectiveness of friends groups in bargaining in the climate change regime. Firstly, the wide diversity of interests relative to climate change among different countries, and the fact that this diversity of interests shifts depending on the specific issue at hand (e.g. LULUCF, financial assistance), makes it difficult for negotiations to be conducted through a coalition, which must pool positions. Only the EU can really negotiate with one voice. The fact that each person at the negotiating table represents a larger negotiating coalition, while reducing the actual number of participants, therefore impinges on that person's ability to make deals or react to developments. Each time a new proposal is presented, or another coalition alters its position, representatives at the negotiating table will have to consult with their constituent coalition before responding. The process can therefore be laborious, as the negotiating coalitions need to meet at each stage before the negotiation can proceed. Interestingly, it was when President

Pronk felt bold enough to ask *individual parties* – rather than coalitions – sitting at the table of the 'Compliance Group' on the last night of COP 6 (part II) whether they agreed with the emerging text that he was able to forge a consensus (see Lefeber, 2001).

Friends groups can also fall in between two stools. That is, they are too small and secretive to be legitimate and fully representative, yet too large and transparent for meaningful negotiation to take place. The exchange of compromises and concessions, with both overt promises and subtle unspoken understandings, is something that simply cannot be done in a room where such a diverse group of political adversaries and allies is present (Yamin and Depledge, 2004). Friends groups also suffer from the mix in participation between ministers and officials, whose differences in approach can make negotiations awkward and unproductive (see also Chapter 13).

Fundamentally, the dilemma lies in the inherent difficulty of limiting participation in a bargaining situation. Delegations, even those that do not participate very actively in open arenas, understandably object to being explicitly excluded, not least because it looks bad for them to have to report to their capitals that they were not present in the key bargaining arena. Moreover, delegations that are not present will not necessarily trust their representative in the friends group to bargain effectively on their behalf and uphold their interests. The transaction costs involved in actually setting up a bargaining friends group can therefore be considerable, and even outweigh the efficiency benefits of the more limited participation. At COP 6, for example, President Pronk spent much of the first week trying to persuade delegations to accept his proposal that the final negotiations in the second week would be conducted in four limited participation groups. This met with such opposition, especially from the G-77, that the plans were shelved, but not before valuable time had been wasted.

There are examples in other regimes where friends groups have been more effective. The negotiations on the Cartagena Protocol on Biosafety, for example, gave rise to the 'Vienna Setting', in which the deal on the Cartagena Protocol was eventually struck (following an earlier breakdown in negotiations). This friends group proved an effective bargaining forum, largely because negotiating coalitions were more cohesive in their interests and positions than in the climate change regime. When President Pronk cited the Vienna Setting as a possible model for organizing COP 6 he received a strong message from parties that it was not appropriate for the climate change regime.

Friends groups in the subsidiary bodies

Given that their conclusions are less politically significant than those of the COP, the SBSTA and SBI Chairs only seldom convene friends groups. There have been several exceptions, however, all with participation by officials, in keeping with the nature of the subsidiary bodies. Friends groups have met where the subsidiary body Chair surmised that the subject matter – usually more technical or procedural issues – lent itself better to consideration in a small, select group, than in a larger, more open forum. 'Friends of the Chair' groups, for example, were convened at SBI 10 to develop draft terms of reference for a workshop on adverse effects, and at SBI 16 to consider a process for the review of the financial mechanism. Friends meetings were also used in the SBSTA as part of the post-Kyoto negotiations on technology transfer. From SBSTA 13, however, these were replaced by open-ended contact group meetings, as there was too much interest in the topic for it to be dealt with exclusively and legitimately in such a small forum. More political friends groups were also in evidence at COP 6, as the Chairs of the subsidiary bodies and open-ended contact groups sought desperately to find ways of reaching agreement.

Shuttle diplomacy

'Shuttle diplomacy' can provide a useful alternative (or complement) to friends groups in the final stages of negotiations. Shuttle diplomacy involves the President (or subsidiary body Chair, or a designated colleague) literally 'shuttling' between each negotiating coalition (and sometimes individual parties), meeting with representatives in succession, and drafting and redrafting text until all the coalitions individually agree to it. Shuttle diplomacy therefore constitutes a form of informal consultation (see above) and, needless to say, presiding officers have always held private meetings on all sorts of issues. This kind of shuttle diplomacy, however, merits separate consideration, given its centrality in sealing a final deal on high profile decisions, including the Bonn Agreements and Marrakesh Accords, along with the New Delhi Declaration. At COP 6 (part II), for example, COP President Pronk met with each coalition following the presentation of his compromise proposal – the 'Core elements' paper – and then with individual Umbrella Group countries to discuss their specific concerns. Similarly, at COP 7, shuttle diplomacy by the ministerial co-facilitators of the 'Fez 1 Group', and by other ministers working on specific issues, was key to reaching agreement on the most contentious points of the Marrakesh Accords, notably how to accommodate the Russian Federation's demands for extra sink credits. Shuttle diplomacy by the COP 8 Presidency was likewise central to agreeing the New Delhi Declaration on Climate Change and Sustainable Development.

The rise of shuttle diplomacy as a means of final deal-making is partly a function of disillusionment with friends groups, and represents a case where the organizers have responded innovatively and pragmatically to problems encountered with more established means of negotiation. The key advantage of shuttle diplomacy is that it allows representatives of negotiating coalitions to meet privately with the President (or other presiding officer), and therefore to talk freely without fear of losing political face, or of being accused of treachery by coalition members. Representatives can speak more openly about the difficulties they face in selling a text to their coalitions, suggesting where movement may be possible, even if it is not yet forthcoming. This would not be possible in a forum such as a friends group, where both coalition members and negotiating adversaries – who could exploit such revelations – are present. It is not insignificant that, at the most sensitive moments of negotiations at COP 6 (part II) and COP 7, the G-77 Chair attended shuttle diplomacy meetings alone. At COP 6 (part II), for example, COP President Pronk received the absolutely critical message during his shuttle diplomacy that each group – bar the Umbrella Group – was prepared to accept his compromise paper as a package, providing all other groups did. Equally important was the message that, although the Umbrella Group had many concerns, it was only the text on compliance that breached the 'bottom line' position of certain members. These messages would probably not have been forthcoming in a more open forum, where the Umbrella Group would most likely have felt compelled to insist on all its demands for much longer, and other coalitions might, in response, have felt compelled to raise their own concerns.

Shuttle diplomacy can also overcome the difficulties in dealing with a mix of ministers and officials, as representatives will meet separately with the COP President (him/herself a minister). Moreover, engaging in shuttle diplomacy is inherently less controversial than convening a friends group, as the right of the presiding officer to meet individually with whatever delegations s/he chooses in the privacy of his/her own office is uncontested. Ironically, however, shuttle diplomacy is even less transparent and procedurally equitable than friends groups, given that meetings are held between just a few people, and the discussions and deals therein are kept secret.

Unofficial negotiations

In addition to the groups discussed above, which operate within the ambit of the official negotiation process, it is common for parties to engage spontaneously in unofficial talks behind the scenes, to bargain and forge deals on issues of particular importance to them. Such unofficial negotiations function *wholly outside the process*, closed to all but the participants themselves, except for the important fact that any results emanating from them may eventually be brought to the plenary or other official forum to be adopted.

Each negotiating session has its own share of unofficial negotiations in the corridors, where individuals exchange ideas and float alternatives on a no-obligation basis, over dinner or a cup of coffee, in the corridors while waiting for an official meeting to start, in the cafeteria queue, or in a privately booked meeting room. Such unofficial exchanges are crucial as lubricants to the official negotiations. They allow delegates to talk completely freely, as individuals rather than representatives of national positions, articulating views and alternatives that would not be possible in an official arena. If necessary, delegates can later deny that any discussion has taken place. Discussions in the corridors can clarify issues or iron out misunderstandings that may have arisen in the official negotiating arenas, while helping to overcome the (usually negative) stereotypes that delegates hold of one another, building up trust and better social relations. It is also through such unofficial conversations that NGO representatives are best able to lobby delegates (see also Chapter 14). Unofficial negotiations thus increase the overall efficiency of the negotiation process, promoting greater progress in the official informal arenas and formal plenaries. It is important not to indulge in a romantic view of such exchanges, however; while they often do encourage a more cooperative atmosphere, highly aggressive bullying also takes place in the corridors.

In addition to routine unofficial talks in the corridors, negotiating finales have often seen more structured unofficial negotiations, whereby the most powerful players have taken matters into their own hands, perhaps frustrated at the constraints imposed by procedural safeguards in the official process, to hold their own deal-making talks. At COP 3, for example, bargaining on emission targets took place principally between the EU, Japan and the US in closed meeting rooms located on an upper floor of the conference centre. The secretariat and presiding officers never observed these negotiations, which have been described as taking place in a 'parallel universe'. For example, while a Swiss delegate was busy conducting informal consultations downstairs on differentiation,[17] the negotiations that would eventually decide this issue were taking place upstairs in the parallel universe. A similar, although more desperate and less successful, round of secret talks took place at COP 6, where top level UK and US representatives gathered in private – while the official friends group was still meeting – to hammer out a deal. This deal, however, was later rejected by the EU as a whole (see also Chapter 12).

Unofficial groups are undoubtedly highly efficient bargaining and deal-making arenas, cutting out all but the most critical players. Participants are able to engage in unrestrained bargaining and full exploration of possible trade-offs, linkages and side payments, without any scrutiny from other parties or NGOs. In the case of COP 3, it is likely that such completely private bargaining was indispensable to reaching agreement on the emission targets of the Kyoto Protocol. However, by their very nature, unofficial talks are not open to collective policy regulation, and are therefore outside any of the procedural equity and transparency safeguards of official forums. Delegates who do not possess the same personal skills, clout or experience cannot engage so effectively in the networks of influence. Moreover, unofficial talks tend to follow, rather than cross, existing lines of political allegiance, particularly between North and South, which creates problems when text eventually emerges into the official arenas.

During the Kyoto Protocol negotiations, for example, countries from the Organisation for Economic Co-operation and Development (OECD) – the most developed countries – had engaged in intensive unofficial talks on emissions trading since roughly COP 2 in 1996, eventually producing a text they had developed – without any developing country input – in the first week of COP 3 (see also Chapter 11). This text, however, was rejected by the G-77, and lingering strong suspicion of emissions trading among developing countries almost blocked agreement on the Kyoto Protocol. In another example, the Russian Federation and Ukraine agreed only to a target in the Protocol of stabilizing their emissions (0 per cent). However, there is evidence that both parties were prepared to accept a –2 per cent reduction target if they had been engaged in negotiations with other parties and pressure had been placed on them. Hungary and Poland then lowered their targets to –6 per cent in protest at the Russian and Ukrainian targets, as well as lack of diplomatic attention from their EU allies (as candidate countries to join the EU, they had been expected to assume the same –8 per cent target as the EU) (Oberthür and Ott, 1999). A non-transparent, non-inclusive negotiation thus probably led to an environmentally weaker outcome than would otherwise have been the case, as well as damaged relations among parties that have continued in the post-Kyoto process.

The case of COP 6 holds similar lessons. Although the deal between the US and the UK was never formally tabled, the fact that it did not even cover financial issues of particular concern to developing countries – because, of course, developing countries were excluded from the negotiations – means that it would very likely have been rejected had it ever reached an official negotiating forum.

The dangers of such explicitly exclusionary negotiations are clear. As a developing country interviewee put it:

> I remember once there was an issue that some Annex I parties they sat together and produced a paper... actually the G-77, I remember, thought the substance was quite OK, but they say they hate this paper. Why? Because they didn't like being excluded... But nobody actually had any problems with the paper itself! All the substance was OK, they just didn't like the way [it was done].

The challenge, therefore, as posed by Freymond, is 'how to accommodate the natural tendencies of the states whose power gives them global responsibilities to settle issues among themselves while also considering the claims of the rest of the international community for participation' (Freymond, 1991, p131). Although nothing can be done to prevent key players from getting together privately to make deals, and indeed such private negotiations are necessary for agreement to be reached, the unofficial negotiations should never be allowed to eclipse the official arenas of the formal plenary, informal arenas and friends groups, if the negotiation process as a whole is to retain legitimacy.

The deal-making arena

Among the network of different arenas upon which negotiations play out, the arena that is chosen (or that simply emerges) as the stage for striking the final deal is of particular importance. The typical pattern in intergovernmental negotiations is for bargaining to become restricted to a smaller and smaller group of parties as the talks become ever more intense with the approach of the deadline. Having started off in a large plenary, negotiations are usually concluded, and a deal reached, in small, closed groups – friends groups, shuttle diplomacy or unofficial talks. Of course, all negotiations will eventually end in a formal, open plenary meeting, but the task of that plenary meeting is typically to rubber-stamp the results of a deal struck in a closed group. At COP 6 (part II), for example, the final deal was struck through shuttle diplomacy by the President, plus a marathon overnight meeting of the 'Compliance Group'. At COP 7, the Marrakesh Accords were similarly agreed through shuttle diplomacy, endorsed by all coalitions in the 'Fez 1 Group'. Other landmark decisions, such as the Geneva Ministerial Declaration at COP 2, the BAPA at COP 4, and the Delhi Declaration at COP 8, were all hammered out in friends groups or through shuttle diplomacy.

In this context, the openness of the finale at COP 3, conducted in a CoW plenary meeting, was remarkable. All parties had the opportunity to contribute to the final approval of each of the Protocol's elements, and putting these together to form an integrated whole, on the basis of full formal procedural equity and transparency. By way of illustration, in the exact reversal of the typical pattern, more interventions were delivered by a greater number of parties on the last night of COP 3 alone than during any full AGBM session, while ten parties made their only formal interventions in the whole Kyoto Protocol negotiation process on the last night.[18] The CoW plenary was fully open also to NGOs and the media, including television cameras and live broadcast on the Internet. There was only one instance where Chair Estrada suspended the open plenary to consult behind the scenes, on the issue of emissions trading where he met privately with the US delegation.

Part of the rationale behind choosing a plenary meeting as the final negotiating arena was precisely its openness to public scrutiny. Estrada wanted to ensure that there would be maximum pressure on negotiators to reach agreement, and that, should any party seek to block consensus, it would be absolutely clear how and on whose responsibility the Protocol had fallen (see also Chapter 14). The choice of a plenary meeting as the stage for a negotiation finale was thus 'rare, but it worked' (interview), achieving efficiency as well as procedural equity and transparency. Two important preconditions, however, to the successful use of an open plenary in this way was firstly the effectiveness in that context of Chair Estrada, and secondly the fact that almost all issues except for the most controversial ones had already been resolved (see also Chapter 12). The final plenary was thus not overburdened and the choices before it were relatively simple, if politically difficult.

Summary and concluding remarks

This chapter has explored the many different arenas upon which the climate change negotiations play out, from formal plenary meetings to unofficial bargaining in cafés and corridors. Each arena fills its own niche in the wider climate change negotiation process. Formal plenary meetings provide a platform for parties to posture if they need to, informal arenas (contact groups, informal consultations) facilitate in-depth bargaining, friends groups can provide advice and insights to the presiding officers and a more intimate bargaining atmosphere, while shuttle diplomacy can be invaluable in allowing completely frank discussion. At the same time, space needs to be given to key players to forge the deals they have to forge in unofficial, private arenas.

The negotiating arena used to work on an issue – plenary, contact group, informal consultation, friends group – often shifts as negotiations progress. Contact groups may break off into informal consultations, informal consultations may expand and formalize into contact groups, or plenary meetings may convene short informal consultations to resolve a tricky outstanding issue.

Each negotiating arena has its own strengths and weaknesses; transparency and procedural equity tends to increase with the formality of the arena, while efficiency generally rises with the degree of informality. Greater efficiency does not necessarily mean greater effectiveness, however. Where procedural equity and transparency have been eroded beyond a certain level of acceptability, then the results of even the most efficient group will not be seen as legitimate.

The various negotiating arenas are thus subject to policy manipulation by the presiding officers and secretariat to exploit their strengths and compensate for their weaknesses. A key tool in this regard is the use of regular plenary meetings to make up for the equity and transparency deficiencies of more informal arenas. It is important, however, not to go too far in this regard. Excessive transparency in the official negotiations can paradoxically prompt the most powerful countries to move behind the scenes to try to strike a deal in the most untransparent way possible.

The key to the effective organization of negotiating arenas thus lies in striking a balance between the different arenas. That is, essentially to achieve an acceptable trade-off between the efficiency necessary to manage complexity and reach agreement on the one hand; and the procedural equity and transparency necessary to secure the legitimacy of the agreement on the other.

Complementary Forums: Workshops, Roundtables and Others

Occasions like that help grease the negotiating cycle (interview).

Introduction

The negotiating arenas discussed in Chapter 9 are complemented in the climate change regime by forums such as workshops and roundtables, which are aimed at gathering information and exchanging views, rather than political bargaining. Such complementary forums are explicitly highly informal, with no set rules governing their structure or conduct of business. Indeed, because complementary forums lack the mandate – indeed, are forbidden – to engage in bargaining, they are less (but not un-) controversial, and therefore the presiding officers and secretariat have even more leeway in their organization. This short chapter explores the various complementary forums that have been used in the climate change regime to 'grease the wheels' of the negotiations proper, focusing on workshops and roundtables.

Workshops and Roundtables

Nature and functions

Like informal groups, complementary forums have been known by different names in the climate change regime. Box 10.1 provides a general explanation of the different terms used, although the use of these has not always been consistent or rigorous.

The Kyoto Protocol and post-Kyoto negotiation processes have seen different trends in the use of complementary forums. During the Kyoto Protocol negotiations, a small number of roundtables and workshops were organized, mostly *during* negotiating sessions, on an ad hoc basis (see Box 10.2 on AGBM roundtables below[1]). In the post-Kyoto period, a new trend emerged of holding workshops *in between* negotiating sessions. As the number of workshops grew, established practices emerged to govern these. Most recently, the subsidiary bodies have begun once again to convene workshops during negotiating sessions, reminiscent of the roundtables held under the AGBM during the Kyoto Protocol negotiations.

The precise aim of complementary forums varies. Some are aimed at exploring and gathering inputs on a certain topic, especially one that is new in the climate change regime. Good examples here[2] are the workshops on the flexibility mechanisms held in April 1999, along with the various workshops held on the adverse effects of response measures. Others have been convened specifically to provide more time to consider lengthy reports. These include the 2000 workshop on the IPCC's Special Report on LULUCF (IPCC, 2000) along with the 2002 workshop on the IPCC's Third Assessment Report (TAR).

Other complementary forums may be focused on supporting national implementation efforts or have a strong capacity-building element. These include workshops to help non-Annex I parties prepare their national communications according to Convention guidelines, along with regional workshops on the implementation of the New Delhi work programme on education, training and public awareness.[3] Still others may be linked more closely to the negotiation process proper, aimed at developing draft text as input for the subsidiary bodies. Workshops on issues related to Articles 5, 7 and 8 of the Kyoto Protocol (methodological issues, reporting and review), for example, along with workshops on LULUCF, became as important to the negotiation process on these topics in the post-Kyoto and post-Marrakesh periods as the negotiations within the subsidiary bodies themselves. In all cases, even in the most technical meetings, it is difficult to prevent political debate, and a degree of negotiation may erupt spontaneously or unofficially on the margins.

In the post-Kyoto era, holding a workshop became an increasingly popular, and relatively uncontroversial, means of kick-starting negotiations on a particular topic (Yamin and Depledge, 2004). In the case of the flexibility mechanisms, for example, the week-long workshops in April 1999 were absolutely critical to improving understanding and ownership of the mechanisms among a wide spectrum of parties, and to identifying issues and options that would need addressing when working on rules for their operation. In some cases, workshops have been held as a means of *avoiding* negotiations, or at least of circumventing the unproductive wrangling that would be generated by attempting to debate controversial issues in a more formal arena. Because delegates are neither negotiating nor making commitments in a workshop, they can discuss a wider range of issues much more freely. The controversy surrounding the issue of policies and measures, for example, meant that no substantive, in-depth debate could take place in the SBSTA itself, and this instead had to be relegated to workshops. The workshop and pre-sessional consultations held on the IPCC TAR provide another example. Open debate on the specifics of the TAR's findings was not possible in the SBSTA itself, due to political sensitivities, especially the fear of developing countries that the TAR's conclusions might be used to call for a new round of negotiations on non-Annex I party commitments. Instead, frank discussion was only possible in the politically more neutral setting of a complementary forum. One interviewee linked the growth in the number of workshops directly to 'the greater rigidity and politicization of

---

Box 10.1 Complementary forums

Roundtables

Several were convened during sessions of the AGBM during the Kyoto Protocol negotiations. *High-level* roundtables, involving ministers, have also been convened during COP sessions (see Chapter 13).

Workshops

A large number of workshops held inter-sessionally have been convened in the post-Kyoto period. The term 'workshop' is politically useful in making it absolutely clear that the forum is entirely informal and technical in nature, and no negotiation whatsoever is to take place. A couple of workshops have been denoted as *expert* workshops, indicating that these are smaller, more select affairs, involving experts on the topic, rather than 'ordinary' delegates. *Regional workshops* have also been held on issues where a regional focus is particularly pertinent, notably technology transfer and capacity-building. A new practice emerging post-Marrakesh is to hold *in-session* workshops.

Expert meetings

A small number of complementary forums have been convened as expert meetings. As with expert workshops, these are smaller occasions, aimed more at providing an opportunity for technical experts to exchange views, ideas and latest research results, than for delegates to engage in discussion.

Pre-sessional consultations

These are held immediately prior to a negotiation session. The term 'consultation' suggests even greater informality than 'workshop', and can also imply stronger political linkage to the negotiation process.

---

almost every dimension of the work of the Convention bodies'.

In many cases, the proliferation of workshops held inter-sessionally is a clear function of the overloaded agenda of the regime bodies, in that there is simply insufficient time during the two-week sessions to devote to in-depth consideration of each agenda item. As one interviewee put it, 'they are an outpouring of what needs to be done'. Another secretariat interviewee explained the multiplication of workshops partly as a fashion: 'there is an element of keeping up with the Jones's. Everyone wants a workshop, even when there's not much to discuss.'

Overall, complementary forums undoubtedly generate more active and open debate than in the negotiating arenas and, as such, contribute to the efficiency of the process. As one interviewee put it, 'they do give a better exchange of views ... so long as they are seen as informal, non-quotable'. Parties are also able to broach politically sensitive topics that cannot be taken up in negotiating arenas, where their mere discussion on the record could imply that they were under formal consideration. Interestingly, one interviewee expressed frustration at this, commenting that 'there's a kind of disconnect between workshops and the formal process'. Although delegates talk more freely in complementary forums, they tend to assume once again the restrictions of their national positions when back behind their country flags. The incisive presentations and free-flowing discussion at the two in-session workshops on adaptation and mitigation held at SBSTA 20 in 2004, for example, contrasted with the rather stagnant proceedings in the formal negotiating arenas at that session.

Characteristics

Complementary forums in the climate change regime share several key characteristics. Most importantly, they are mandated by one of the regime bodies (usually one of the subsidiary bodies), and explicitly designated as technical forums where no bargaining is to take place – only information gathering and sharing (Yamin and Depledge, 2004). During the Kyoto Protocol negotiations, parties were content for complementary forums – notably the AGBM roundtables – to be convened at the initiative of the Chair according to a relatively loose mandate. Post-Kyoto, however, as the number of workshops has multiplied and their prominence in the negotiations has increased, parties have become more anxious to exert greater control over their organization, and to specify a more detailed mandate, in some cases even an agenda and terms of reference, especially for workshops on more controversial topics.

Funding for workshops is provided through the trust fund for supplementary activities, rather than the core budget, which means that it is reliant on voluntary, ad hoc donations from parties. Funding will typically be provided by (usually Annex II) parties particularly interested in the topic under discussion (Yamin and Depledge, 2004). Denmark and France, for example, funded workshops on policies and measures, while Canada, Germany, Finland and Japan jointly sponsored the 2003 LULUCF workshop held in Brazil. The growing number of workshops is leading to what one interviewee called 'workshop fatigue' among donors and therefore difficulty in securing adequate funds for each meeting.

Inter-sessional workshops are held all around the world, as countries offer to host a workshop on an issue of special relevance to them. The Russian Federation, for example, hosted a workshop on joint implementation projects in May 2004, while Iran (an OPEC country) hosted a workshop on economic diversification in 2003. In some cases this provides an opportunity for countries who could not afford – in financial or institutional terms – to host a full COP to play host to a part of the climate change regime process. This contributes to spreading a sense of ownership of the climate change negotiations more widely, which can be especially important in the developing world. A workshop held in Mauritius, for example, was opened by the Prime Minister himself, illustrating the value placed by that country on hosting such an event. Where there is no offer to host, the meeting will be held in Bonn, the secretariat headquarters, to minimize costs.

A practice that has emerged to cope with the growing volume of workshops is to hold these in the few days immediately prior to a negotiating session and in the same venue, in order to save on cost (such workshops are then often termed 'pre-sessional consultations' – see Box 10.1). This practice, however, while more cost-efficient, does mean that key parties and individuals who tend to get invited to many workshops are expected to be away from home for up to three weeks at a time (Yamin and Depledge, 2004). One interviewee, an NGO delegate, complained that, although his constituency had been invited to pre-sessional consultations, he had not been able to find any representative willing to attend, as the time away from home was too long. Such concerns have prompted the re-emergence of workshops held during negotiating sessions.

Box 10.2 The AGBM roundtables

The AGBM roundtables were a 'procedural innovation' (interview) at the time. They were particularly in evidence at AGBM 4 (July 1996), where three were held, on 'Quantified emission limitation and reduction objectives – QELROs' (the AGBM euphemism for emission targets), 'Policies and measures' and 'Possible impacts on non-Annex I parties of new commitments for Annex I parties'. Anyone was permitted to attend, although the small size of the room restricted participation. Each roundtable benefited from a panel of speakers from parties, IGOs and NGOs, whose presentations were followed by open discussion. The first two roundtables were chaired by members of the AGBM Bureau, the third (unusually for the climate change regime) was chaired by a delegate from a research NGO. Each Chair presented a report on the roundtable to the AGBM plenary, which were annexed to the formal AGBM report. Interviewees stated that they had found the roundtables helpful. One commented: 'They were a good innovation...like the one on impacts on developing countries...some delegations were able to say in a much more open way what their real feelings were'.

Workshops are often chaired by the Chair of the relevant subsidiary body, although s/he may invite another delegate – perhaps an expert on the topic – to chair on his/her behalf. The 2000 LULUCF workshop held in Italy, for example, was co-chaired by an Annex I and non-Annex I representative, who co-chaired the contact group on this issue in the formal negotiations. Complementary forums often include presentations from experts, including from the NGO and IGO communities. Interestingly, some secretariat interviewees reported that the large number of workshops is starting to exhaust the supply of qualified NGO/IGO 'resource persons'.

The AGBM roundtables held during the Kyoto Protocol negotiation process were open to participants from all parties, and also to observers. Participation at the post-Kyoto workshops, however, which have almost all been held inter-sessionally, has been on an invitation-only basis. This is necessary for logistical reasons, while also promoting the efficiency of discussions through a more intimate and expert atmosphere. Moreover, the cost of attending such inter-sessional meetings inevitably deters widespread attendance, especially from developing countries and EITs. The number of participants varies from workshop to workshop. The larger workshops of more general interest have reached participation levels of 100+, while expert workshops/meetings, or workshops focused on national capacity-building and training, have been smaller affairs, with 30–50 participants. Invitations are typically issued to countries, who then nominate their own experts to attend. Funding is provided to cover the costs of participation by eligible parties, but other invitees must cover their own costs. The invitation list is drawn up by the Chair of the workshop, along with the presiding officer of the convening body (often one and the same), with the secretariat's help.

A limited number of NGO representatives are usually invited to observe proceedings, or indeed to contribute through presentations and commentary. Invitations are typically issued through the NGO constituency focal points (Yamin and Depledge, 2004). NGO participation at workshops became a bone of contention in the post-Kyoto period as workshops grew in importance and NGOs were concerned that their attendance was being excessively constrained. Over time, however, and with prompting from the SBI, the secretariat has developed practices that now ensure that the number of invitations generally exceed the number of participants. For more on NGO participation, see Chapter 14.

The limited participation nature of the post-Kyoto workshops introduces an element of controversy to them. With only a subset of parties and NGOs present, transparency and procedural equity are inevitably affected. As one NGO interviewee put it, 'only the resource rich delegations can send somebody and the NGOs are de facto excluded. We have no funding to go. The rich NGOs may be able to send one or two, but you have a very rarefied atmosphere.' The acceptability of workshops is therefore dependent on the representativeness of participation, making available as much information about them as possible, and the strict upholding of their status as complementary – rather than negotiation – forums. To enhance transparency, workshop presentations and other documents are posted on the UNFCCC website (www.unfccc.int), while the Earth Negotiations Bulletin (ENB), an independent reporting agency that reports on formal negotiating sessions, has been commissioned to report on major workshops of general interest, where funding is available.[4] The convening of in-session workshops in the post-Marrakesh period, where attendance is open to all, is at least a partial response to concerns over the necessarily limited participation in inter-sessional workshops.

The outputs of complementary forums have varied. Some workshops in the post-Kyoto era have resulted in just an oral report, without written record (e.g. the 1999 workshops on the flexibility mechanisms). More recently, the trend has been to issue a written report by the Chair as an official document (e.g. the two workshops on policies and measures and the 2002 IPCC TAR workshop). Alternatively, where it was deemed preferable not to give a workshop report such official status, the oral report by the Chair has simply been placed on the UNFCCC website (e.g. the in-session workshops on adaptation/mitigation at SBSTA 20 in 2004). Other workshops have actually produced draft text, which has then been used as input for discussions at the following formal negotiating session. The 2001 workshop on Kyoto Protocol

Articles 5, 7 and 8, for example, produced a very important revised negotiating text on these articles, which was subsequently used in the negotiations at COP 7. In this respect, some workshops have provoked concerns that they have strayed too far towards negotiation, and in certain cases their outputs have been disputed by the subsequent formal negotiating session. The detailed work programme for including LULUCF activities under the CDM prepared by the 2002 workshop, for example, was challenged at the following session of the SBSTA, notably by the G-77, which expressed concern at the limited nature of participation at the workshop (see ENB, 2002).

An interesting dimension to complementary forums is that delegates will often feel able to debate more freely, despite the retention of many of the features of informal contact groups or even formal plenary meetings, including, in some cases, interpretation, sound recordings and presence of NGOs. Where a written report is produced, these are ironically much more detailed than the reports on proceedings in the formal negotiations themselves. This reveals the important distinction in multilateral negotiations between negotiating and non-negotiating arenas – in addition to the distinction between formal and informal settings; far more personal and radical views can be expressed in a forum that is explicitly designated as being not for negotiation, almost irrespective of the procedures in place.

Side events and ad hoc meetings

In addition to the workshops and roundtables discussed above, *side events* also make up an important set of complementary fora. While these are mostly organized by NGOs (and are therefore discussed in Chapter 14), the subsidiary bodies, secretariat and parties themselves make use of side events as highly informal arenas where issues can be discussed in a politically safe environment.

Side events on the research recommendations of the IPCC TAR, for example, involving representatives of research organizations, were held under the auspices of the SBSTA in 2002 and 2004. These events were useful in providing more time for discussion than would have been available in the formal negotiations, while enabling parties and researchers to engage in a real discussion on research needs outside the onerous constraints of political acceptability of the negotiating arenas.

Side events organized by the secretariat where non-Annex I parties present their recently submitted national communications have now become a regular fixture of the climate change regime calendar. These events make an important contribution to the exchange of information, especially among developing countries on what works in their particular context. Due to the controversy that surrounds developing country actions under the climate change regime, these side events are one of the few arenas where the climate change policies of developing countries can be presented and discussed in an open manner. A further initiative by the secretariat has been the organization of themed 'high level' side events, often involving partnerships with business groups, to engage in informal discussion on specific aspects of the international response to climate change (see also Chapter 6).

Parties make use of side events to showcase particular aspects of their climate change policy and programmes, to explain the expected impacts of climate change on their territories, or even to present new proposals and ideas. During the Kyoto Protocol negotiations, for example, the EU convened a side event at AGBM 5 in late 1996 to present and discuss its proposals in more detail, in response to concerns and misunderstandings expressed in the formal negotiations. In another interesting example, the US took advantage of the informality – yet high visibility – of the side event schedule at COP 9 in 2003 to organize two events on its climate change policies. This sought to convey the politically significant message that the US still takes (or wishes to be seen to take) climate change seriously, despite its repudiation of the Kyoto Protocol. Interestingly, as also discussed in Chapter 14, an increasing proportion of side events are being organized by parties.

Summary and concluding remarks

This chapter has discussed how complementary forums, while not central to the negotiation process, are certainly important accompaniments to it, performing a variety of roles that help to increase the efficiency of the negotiations. Central to their role is the promotion of more candid debate among delegates, outside the strictures of negotiating arenas. A remarkable, and somewhat troubling, aspect of the climate change regime is indeed how, in many cases, complementary forums have become arenas not only for more in-depth discussion, but also for broaching topics that would be vetoed in the negotiating arenas. This is useful, but can also lead to frustration when constructive discussions in the complementary forums fail to generate discernible change in the positions of parties in the negotiations proper. As non-bargaining arenas, roundtables, workshops and similar events can never actually reconcile the positions of parties. Instead, at their best, they can help pave the way for more effective bargaining by facilitating the flow of information and ideas.

Texts

The basic work of the negotiations...to produce texts that translate concepts into words (secretariat interviewee).

Introduction

The story of the Kyoto Protocol negotiations is essentially the story of how over 430 pages of written proposals by parties gradually evolved into the 25 page final, authentic text of the Kyoto Protocol that now resides at UN Headquarters in New York. The story of the post-Kyoto negotiations is a similar one, albeit more complex and lengthy, with over 2000 pages of proposals eventually, and laboriously, resulting in the 23 decisions covering over 210 pages of the Marrakesh Accords. Indeed any negotiation – from the high-profile Kyoto Protocol and Marrakesh Accords to the most humble subsidiary body conclusions – can be characterized as a process of struggle between parties to secure the translation of their favoured ideas into texts. This struggle is mediated by the organizers of the negotiation process who seek words, concepts and phrases that can cover enough differing preferred concepts to enjoy general consensus among parties. Texts thus play a crucial role in facilitating and concretizing the reaching of agreement among parties. As a secretariat official once put it, 'any discussion without text is almost a waste of time because, when confronted with text, theoretical convergence [among parties] turns into real divergence'.

Many different kinds of documents are produced in the climate change regime, including provisional agendas and annotations, reports, and various background information documents. In this chapter, we concern ourselves chiefly with the documents at the centre of the negotiation process, namely, *single negotiating texts*, including *Chair's texts*. We also examine the textual building blocks through which single negotiating texts are constructed, namely *raw material* texts (proposals submitted by parties) and *precursor* texts (preludes to negotiating texts that compile proposals or provide an inventory of options). After explaining general issues relating to documentation, this chapter examines the textual development process that has unfolded in the climate change negotiations, from the raw material submitted by parties through to the final agreed written output of decisions or conclusions.

Documentation in the climate change regime

The climate change regime, like all bodies in the UN system, works through various different types of official documents, each with their own symbols and indeed symbolic meaning. These official documents of the climate change regime are processed by the UN office at Geneva, and then archived in the UN Library. Although most of the main document types – regular documents, INFs, MISCs, CRPs and others – are prevalent throughout the UN system, the climate change regime has, within the bounds of UN document processing rules, developed its own practices for using them, and has assigned different connotations to them. The regime has also developed a couple of its own document types – notably the WEB series – reflecting its own unique needs and circumstances. The main official and semi-official document types in the climate change regime, their symbols and key uses, are explained in Table 11.1. These are not just archaic, bureaucratic classifications. How a text is published – as a regular, INF, CRP or L document – carries with it important connotations in terms of its status and intended use that can have significant implications for its acceptability and prospects in the negotiations. An appreciation of the different meanings of these document types is critical to understanding how texts have evolved in the climate change negotiation process.

Within the bounds of established practices summarized in Table 11.1, there is considerable leeway for the organizers of the negotiation process to exercise judgment in how texts are published in order to advance the negotiations. During the Kyoto Protocol negotiations, for example, Chair Estrada published the negotiating text that came out of AGBM 7 in mid-1997 as an INF document in English only, but his own Chair's text as a regular document in all UN languages. This sent a very clear signal that, at the forthcoming AGBM 8 session, negotiations should centre on the Chair's text, and not the INF document with its much lower status.

In addition to the official and semi-official documents explained in Table 11.1, the climate change negotiations make extensive use of *non-papers*. These are documents that have no official status, and therefore no symbol or even secretariat logo. They are simply photocopied on blank paper. Non-papers are not usually placed on the secretariat website and, except for non-papers of major importance that are kept by the secretariat (but not the UN) library, they are not formally archived in any way. Non-papers are absolutely crucial to the negotiation process for two reasons. Firstly, because they are not subject to official processing, they can be photocopied and circulated very quickly. This can be very important in accelerating negotiations when time is short, or in capturing emerging, fragile agreement. Secondly, because non-papers have no official status, they can be used much more freely to explore different proposals and possible compromises. In a delicate negotiation, parties can refuse to allow proposals they object to even to be circulated in an official document, for fear this will lend them some kind of standing. Circulation in a non-paper, however, implies no status or commitment, and is therefore much more acceptable. In this sense, non-papers are the documentation equivalent of informal negotiating arenas, or even complementary forums (see Chapters 9 and 10). A good example of this point occurred in the final stages of the Kyoto Protocol negotiations at COP 3, where Annex I parties were engaged in unofficial talks behind the scenes on a text on emissions trading. The G-77, which had not been included in those talks, conceded to the circulation of that text *as a non-paper*, but Chair Estrada subsequently refused to include it in the official single negotiating text, knowing that the developing countries would object to endowing the text with official status in this way.

In all cases, in keeping with the close working relationship and division of labour between presiding officers and the secretariat (see Chapter 6), it is typically the secretariat who will carry out the bulk of the work in drafting texts – whether precursor texts, negotiating texts, or even Chair's texts. The presiding officer will usually review the secretariat's work, maybe propose changes, and then present the text under his/her own responsibility.

The development of texts in the climate change negotiations

No specific rules exist to govern how texts gradually develop in a negotiating round under the climate change regime. Established practice, however, is broadly for a negotiation to pass through the stages of textual development outlined in Table 11.2 below, corresponding to the main stages of a negotiation described in Chapter 2:

The cardinal law of any textual development process is that it must always keep moving, thereby both reflecting and prompting the progress of negotiations from exploration and the tabling of proposals, to bargaining, to deal-making. If a textual development process gets stuck at any stage – publishing endless raw material from parties, say, or stubbornly lengthy negotiating texts session after session – this not only suggests that the negotiation is stagnating, but also perpetuates that stagnation, as weighty, complex documents raise obstacles to engaging in effective bargaining or deal-making.

We now examine various cases of textual development in the climate change negotiations, focusing on the Kyoto and post-Kyoto negotiation processes. In doing so, it is important to bear in mind the differences between these two sets of negotiating rounds, notably the greater disaggregation of the latter. In the Kyoto negotiations, all texts – from the initial raw material from parties to the final negotiating texts in Kyoto – covered the whole package of issues under negotiation (emission targets, policies and measures, compliance, reporting, and so on). In the post-Kyoto negotiations, however, the development of texts on each issue followed its own separate path. This chapter focuses in particular on the flexibility mechanisms, compliance and LULUCF, which were each dealt with in separate negotiating arenas (see Chapter 9) with a separate textual development process. It was only at COP 6 that the key political issues – the

Table 11.1 *Official and semi-official documentation of the climate change regime*

Document	Symbol	Typical content, use and connotation	Language
Regular	No special suffix e.g. FCCC/CP/2002/1	The most formal document type. Used for reports on negotiating sessions (including conclusions and decisions), agendas, most secretariat background documents	All six UN languages and subject to rules on advance distribution
Information	INF suffix e.g. FCCC/SBI/2002/INF.2	An informal document. Used to publish technical data and information (e.g. a scoping study), or workshop reports. Documents that should be issued as regular documents are sometimes published as INFs simply because there is insufficient time to translate them	English only
Miscellaneous	MISC suffix e.g. FCCC/SB/1998/MISC.1	Informal document, compiling proposals or views submitted by parties	Language of submission. No formal editing
Technical paper	TP suffix e.g. FCCC/TP/2001/2	Informal document, used for detailed analytical papers on technical issues. In theory, not linked to any specific negotiating session, but often used as background	English
Limited distribution	L suffix e.g. FCCC/CP/2003/L.2	Formal document prepared in-session (during negotiating sessions) containing draft text (decisions or conclusions) presented to the COP or subsidiary bodies for adoption. Distribution is limited in that only a small number of copies are printed, but not restricted	All six UN languages
Conference room paper	CRP suffix e.g. FCCC/SBI/1999/CRP.3	Informal document prepared in-session containing proposals or text reflecting the status of negotiations on a particular issue. Draft text presented to the subsidiary bodies or (less often) the COP for adoption that should be issued as an L document are sometimes published as CRPs due to lack of time for translation	English only
Web-only	WEB suffix e.g. FCCC/WEB/2002/2	A new series of only semi-official documents published only on the secretariat website without going through UN channels. Web documents are a pragmatic response of the secretariat to growing demand for documents and the need to minimize costs. Used for material that is regularly updated or does not require official status, e.g. submissions from NGOs and updates on the preparation of non-Annex I national communications	English. No secretariat logo

Source: Table derived from FCCC (2000) and Yamin and Depledge (2004)

Table 11.2 *The textual development process*

Negotiating phase	Typical text
Exploratory phase and tabling of proposals	The submission of proposals (raw material) from parties and their publication in *MISC documents*. This may continue throughout the negotiation, but with the volume of submissions and new ideas contained therein tailing off
Transition towards bargaining	One or two *precursor texts* (prelude to a negotiating text), either an inventory of options or a compilation of proposals
Bargaining phase	An initial *single negotiating text*, followed by a series of increasingly streamlined revisions
Intensive bargaining and start of deal-making	A *Chair's text*, presenting (elements of) a (more or less comprehensive) compromise
Deal-making	A final round of *single negotiating texts*, probably focusing on key outstanding issues and including Chair's proposals

'crunch issues' – were brought together for consideration as a package deal, eventually resulting, at COP 6 (part II), in the Bonn Agreements.

Another structural difference between the two processes was that, because the Kyoto negotiations were mandated to adopt a new treaty – a protocol or an amendment – they were subject to the formal deadline of circulating a text at least six months before COP 3. Supported by a legal opinion from the UN Office of Legal Affairs, this provision was interpreted by the organizers, and eventually endorsed by the parties, as referring to the circulation of a formal negotiating text, after which point no more 'substantively new elements' could be put forward (AGBM 6 report, p6; see also Chapters 5 and 7). This deadline – 1 June 1997 – proved very helpful to the textual development process, as discussed further below. The post-Kyoto negotiations, however, which were to result only in COP decisions, had no such formal deadline, which again affected the textual development process.

Each step in the textual development process of the Kyoto and post-Kyoto negotiations is explored in more detail below, and summarized in Tables 11.3 to 11.8.

The raw material: miscellaneous documents

The basic documentary components for the climate change negotiations are the textual proposals put forward by parties and issued in miscellaneous – MISC – documents. These written inputs are submitted in response to requests from the subsidiary bodies or, less often, the COP. Because MISC documents simply reproduce parties' proposals verbatim, they are routinely used as an uncontroversial means of launching or advancing a negotiation process. Any new negotiating round will pursue its initial exploratory stage with a call for proposals by parties and their issuance in MISC documents. This applies both to major negotiating rounds, and to the more routine consideration of any new issues, or new aspects of issues, in the subsidiary bodies.

The Kyoto Protocol negotiations began with two rounds of party submissions, circulated in MISC documents. The first round of submissions, up to and including AGBM 3 in early 1996, consisted of 'comments' by parties. At AGBM 3, however, this request for submissions was upgraded to call for 'proposals', and at AGBM 4 in mid-1996 for '*concrete* proposals', thereby encouraging parties to propose more specific draft text in legal language, rather than a vague exposé of ideas. The six-month deadline for the circulation of the formal negotiating text proved very helpful in providing a clear watershed, after which time parties were not expected to submit major new proposals. Therefore, although submissions continued to be received by the secretariat and issued in MISC documents after the six-month deadline had passed on 1 June 1997, these were largely elaborations on previous proposals, and of lesser page length. In the six months between 1 June 1997 and the start of COP 3, for example, 63 pages of MISC documents were produced, compared with 267 the previous six months.

MISC documents were particularly important for the post-Kyoto negotiations, given the novelty and unprecedented nature of the issues – notably the flexibility mechanisms – under negotiation. The opportunity to communicate their views in MISC documents was crucial in providing an outlet for parties to reflect on the new issues arising from the Kyoto Protocol. Altogether, in the two years before scheduled deadline for the post-Kyoto negotiations at COP 6 in November 2000, over 740 pages of MISCs were submitted on the flexibility mechanisms alone, and over 630 on LULUCF. The absence of any formal six-month deadline for a negotiating text akin to that for the Kyoto Protocol negotiations meant that the volume of MISC submissions continued at a high level for longer. In the case of the flexibility mechanisms, for example, 164 pages were received just two months before COP 6. Such a massive volume of written submissions at such a late stage inevitably complicated the negotiations. The situation for LULUCF appeared even more dramatic, with 450 pages submitted between August 2000 and COP 6. This late barrage of submissions was due to the structure of negotiations on LULUCF; much of the substantive consideration of the topic was delayed until after the publication of the IPCC Special Report on LULUCF (IPCC, 2000) in May 2000 (see Chapter 12).[1]

The main function of MISC documents, as identified by the former Executive Secretary, is to act as 'a vehicle for sharing ideas with other parties' (Zammit Cutajar, 1995). MISC documents are the recognized channel for governments to publish and circulate their official views and proposals, and are made widely available to all parties, observers and the global public. From 1997, most of the contents of MISC documents have been placed on the secretariat website (www.unfccc.int) widening their accessibility.

MISC documents also play a crucial role in reassuring delegates that their views are on the table and under official consideration. The inclusion of proposals in an official MISC document, with its own document symbol and the Convention logo, confers recognition and status upon a party's written text, which then resides in UN documentation archives as a permanent textual record of a party's position. It is more important for some parties than others to gain official recognition for their proposals in this way. For example, during the Kyoto Protocol negotiation process, both the G-77 and China, and the US, announced their proposed emission targets at AGBM 8. While the G-77 and China requested that their proposal be included in a MISC document (see FCCC/AGBM/1997/MISC.8), the US never did so, and no official record of the proposed US target therefore exists within the regime. This can be partly attributed to the confident assumption of the US that its proposal would be given due attention in the negotiations, whereas the G-77 and China, a group of developing countries aware of their more limited power, sought the official backing and recognition of a MISC document. Referring to the G-77 and China proposal at AGBM 8, Mwandosya (then Chair of the G-77) tellingly remarks, 'for appearing in *an official document* the matter was *formally on the table*' (Mwandosya, 2000, p85, emphasis added). This tendency has continued in the climate change regime, with the G-77 and China often insisting that their statements made during negotiating sessions be issued in MISC documents.

Once a single negotiating text has emerged in the later stages of negotiations, MISCs can be used by the organizers of the negotiation process as a form of reassurance, to bolster the confidence of parties that their proposals are still under consideration, even if their contents are not fully covered in the single negotiating text. During the Kyoto Protocol negotiations, for example, Chair Estrada qualified his presentation of the initial single negotiating text by saying, 'all...the various miscellaneous documents...remain on the table' (AGBM 7, 1997c). He made similar remarks at later stages in the negotiations (AGBM 8, 1997b; CoW, 1997a). Although negotiations in fact focused on the relevant single negotiating texts and the MISC documents were scarcely referred to, this reassurance helped to increase the acceptability of those texts. In this sense, MISCs can be important as *face-saving* tools. Delegates who fail to get their chosen concepts reflected in a negotiating text can claim, in particular to their domestic constituencies, that their proposals are still on the table. During the Kyoto Protocol negotiations, for example, the US, in accepting to work with the Chair's text presented in late 1997, pointed to the introductory statement in that text, to the effect that it was presented without prejudice to the original proposals from parties contained in the relevant miscellaneous documents (AGBM 8, 1997a). This enabled the US to accept the Chair's text as a basis for negotiation, even though it did not contain the highly controversial US proposal on the evolution of commitments for developing countries.

Precursor texts

The term 'precursor texts' is used here to refer to texts that group and organize proposals from parties into a single document, but are not yet fully-fledged negotiating texts. The relevance and importance of precursor texts lies in the fact that the designation of a document as a 'negotiating text' can be an extremely important symbolic moment in a negotiation process. Parties will often be hesitant to declare that they are engaging in actual bargaining – which is implied when a formal negotiating text is on the table – preferring to stick to exploratory work and the tabling of proposals for as long as possible. This was the case in the Kyoto Protocol negotiations, for example, where there was great sensitivity over when real negotiations would commence. Precursor texts can therefore serve a useful purpose in advancing the consideration and textual development of an issue – organizing submissions under common headings, grouping similar proposals together, reducing duplication in the texts, identifying key issues, pinpointing areas where proposals are lacking – but stopping short of taking the politically significant step of publishing an initial negotiating text. In this sense, precursor texts often mark, and encourage, the transition from the exploratory stage of negotiations and the tabling of proposals, to the start of bargaining.

This begs the question of what distinguishes a negotiating text from a precursor text. Often, the distinction is simply in the name given to the document. There are, however, certain 'markers' that indicate a true negotiating text (see below) and these, therefore, will usually be absent from a precursor text. A typical precursor text will thus:

- be drafted at least partly in a narrative style, with only limited legal text
- retain the separateness of individual proposals, often simply listing these rather than integrating or consolidating them into common text
- attribute text to the party that proposed it and
- not include square brackets, which indicate text that is not agreed and are classic markers of a negotiating text.

In some cases, however, as the presiding officers and secretariat seek to advance the negotiations, some of the markers of a negotiating text – absence of attribution to the proposing party, for example, or inclusion of square brackets – may creep in. Indeed, smoothing over the pathway from raw material to a negotiating text is one of the main functions of a precursor text.

Single negotiating texts

The literature assigns an important role to single negotiating texts, viewing these as key tools for promoting a more cooperative bargaining approach and facilitating agreement. Single negotiating texts are thought to encourage opposing parties to focus on text rather than on the differences between them, and to allow many issues to be considered simultaneously, thus facilitating trade-offs and enhancing the efficiency of the process. According to Fisher et al (1999, p122):

> The one-text [single negotiating text] procedure ... is almost essential for large multilateral negotiations. One hundred and fifty nations ... cannot constructively discuss a hundred and fifty

different proposals. Nor can they make concessions contingent upon mutual concessions by everybody else. They need some way to simplify the process of decision-making. The one-text procedure serves that purpose.

A negotiating text is basically a draft of the output (protocol, decision, conclusion) under negotiation, with language that is not agreed appearing in square brackets, or otherwise highlighted. The aim of the negotiation is thus to work on that outstanding language, finding compromise text that can lift – or remove – the brackets.

As noted above, producing an initial negotiating text can be a very important step in a negotiation process, both reflecting and facilitating the start of bargaining. Options are gradually whittled away and discarded, rather than just consolidated and rearranged. The contents and structure of the initial negotiating text can therefore be very politically sensitive. Whatever the structure of the negotiating text, it is likely to persist to a large extent through to the final agreement. If a proposal appears in the negotiating text, then it is formally under consideration. Conversely, if it is not covered by the negotiating text, then it is unlikely to be given serious attention.

The sensitivity surrounding the preparation of a negotiating text can be illustrated by the realization that the term 'negotiating text' is in fact rarely used. Even the negotiating text issued according to the formal six-month deadline in the Kyoto Protocol negotiations was not titled as such (see Table 11.3). Neither the post-Kyoto compliance negotiations nor the LULUCF negotiations ever brought forth a document with the title 'negotiating text', preferring to use more neutral, safer titles (see Tables 11.6 and 11.7).

Markers of a negotiating text

Although there are no rules governing the structure, style or contents of a negotiating text, there are three main markers for such a text. Firstly, the key marker of a negotiating text is that the proposals from parties are no longer distinct and attributed to their proponents, but are merged, consolidated and integrated into common text. Removing the names of proposing parties is particularly important for bargaining to start. Without attribution, the text is less personalized, and it is therefore easier for parties to consent to the deletion of what is no longer 'their' text.

Secondly, divergent preferences are indicated by way of square brackets. Square brackets are indeed the standard means in intergovernmental negotiations of signifying lack of agreement over a particular piece of text. However, there are instances where the organizers of the negotiation process might wish to convey a more subtle nuance to a phrase or paragraph beyond stark agreement or disagreement. In such cases, differing textual devices may be used, such as footnotes or notes to the reader. Other more innovative textual devices have been used in the climate change negotiations to cope with and reflect the great complexity of texts, made possible by modern word processing. At COP 3, for example, side bars in the document margins were used to indicate those articles still requiring political negotiation in the text produced for the arrival of ministers (see Table 11.3). Such side bars were used in preference to square brackets, which Chair Estrada sought to avoid, given their symbolism of disagreement among parties. The post-Kyoto negotiations on the flexibility mechanisms were particularly innovative in their use of textual devices. The negotiating text emerging from COP 6 (part II), for example (see Table 11.5), used '+' signs to indicate the status of various parts of the text. '+++' was used to denote text adopted as part of the Bonn Agreements, '++' for text agreed in drafting groups and '+' for text partially agreed in drafting groups, with unmarked text either not considered or not agreed. This notation was extremely useful for parties to make sense of the very complex outcome of COP 6 (part II), and to help them embark on the final stages of negotiations at COP 7 with a good understanding of the status of negotiations. The simple use of the traditional square brackets could not have coped with the complexity of the text.

Thirdly, and very importantly, negotiating texts are recognized as such primarily by their use of legal language. In a legal agreement, minute details and nuances can make all the difference. It can often prove relatively simple for parties to agree on broad conceptual ideas, but much more difficult to agree on transcribing and codifying that conceptual concurrence into legal text. Legal text consists of a common language whose subtleties, intricacies and 'climate-specific' jargon are understood by all experienced negotiators. 'Shall', 'should', 'must', 'may', 'might', 'will' are among the more obvious and common terms in negotiations, each with their own implications that can make or break an emerging compromise.

The importance of working on text in legal language can be illustrated by the experience at COP 6. Here, COP President Pronk adopted a different approach based on the assumption that ministers would prefer to negotiate on political concepts and broad-brush deals, not commas and brackets. The suggestion was that, once the political deals had been struck, it would then be possible for the secretariat or technical negotiators to write these up into legal text. The effect was that the negotiating texts, which had developed over two or three years, were sidelined, as negotiators groups were encouraged to focus instead on negotiating political deals. This approach, however, proved ineffective. The negotiation Chairs repeatedly complained that, whenever they thought they were moving forwards in their political negotiations and sought to concretize the concepts discussed on a piece of paper, the apparent agreement tended to evaporate.

Negotiating texts as codifiers of agreement

It is common for several iterations of a full negotiating text to be produced under a negotiating round. Typically, a new negotiating text will emerge from each negotiating session, reflecting the status of discussions at the close of that session. Then the presiding officers will typically seek a mandate to work further on that text, with the help of the secretariat, for the following session. During the negotiation finale – the final negotiating session – it is typical for two or three iterations of the full text to appear as official conference room papers (CRPs), along with myriad iterations of specific parts of the text appearing as unofficial non-papers. Regular revisions of a negotiating text are extremely important as a means of concretizing fragile, emerging agreements, turning tentative compromises into reality. Once a text has been codified in writing, it tends to become permanent, especially if aided by strong chairing. During the COP 3 finale of the Kyoto Protocol negotiations, for example, unbracketed texts on several issues (e.g. review of commitments, provisions for EITs) were included in negotiating texts on a preliminary basis, but subsequently remained unchanged. Conceptual agreement and text became one and the same, so that the emerging agreement was built up and solidified organically through a bottom-up process – word by word, comma by comma – and could not then be easily unravelled.

The first couple of iterations of a negotiating text are almost always lengthy, as more proposals are included than will finally be adopted, and parties are anxious to have their views reflected in the text. As one interviewee put it, '. . . all negotiating texts . . . are a kind of a dumping ground for 150 countries' ideas. [They] tend to develop elephantitis.' The key to a successful negotiation is for the negotiating text to be gradually streamlined and simplified with each revision, as agreement is reached on parts of the text and options are taken off the table. The initial negotiating text for the Kyoto Protocol negotiations, for example, was very unwieldy, at 129 pages long. This was cut down to 82 pages at AGBM 7 in mid-1997, and to 32 pages by the start of COP 3. This negotiating text was then whittled down to 24 pages by the second week (see also Table 11.3). It was not only the number of pages that was reduced, but also the number of outstanding issues reflected in the text. The 94 pairs of brackets appearing in the revised negotiating text at the start of COP 3 were reduced to 81 a week later. No brackets at all were included in subsequent iterations because the Chair sought to put forward compromise proposals, but the number of areas of disagreement was also reduced. The post-Kyoto negotiations on compliance and LULUCF also followed a similar process of gradual consolidation and streamlining of their negotiating texts (see Tables 11.6 and 11.7).

The negotiating text for the post-Kyoto negotiations on the flexibility mechanisms, however, developed in a dysfunctional manner, even considering the large number of complex issues that it had to cover.[3] Despite numerable iterations (see Table 11.5), the document showed little appreciable, sustained streamlining or slimming down until COP 7 itself. As the Earth Negotiations Bulletin put it, in their analysis of SBSTA/SBI 13, 'Progress on mechanisms was disappointing. Parties arrived . . . with a 125 page text, and departed with a 200 page text and an assurance that they will be able to make further submissions prior to COP-6' (ENB, 2000d). This was due to two factors: firstly, parties felt the need to keep putting forward written submissions and amendments, so that the negotiating text just kept on growing. Even at COP 6 (part I), some 50 pages of amendments were submitted to the text that appeared at the close of the first week of negotiations. This was, in part, due to the legitimate need of governments to gain ownership over the three mechanisms, especially the CDM, which, in Kyoto, had been developed by only a small number of parties.

However, the lack of restraint displayed by parties in putting forward new submissions and amendments took this legitimate need to unreasonable extremes. This tendency was not helped by the second factor, that is, the approach of the Chair of the joint contact group on mechanisms, who adopted a liberal attitude of encouraging buy-in, rather than a sterner approach that might have put a stop to so many submissions. This cautious strategy may have been appropriate given the controversy surrounding the flexibility mechanisms, but it did not help to produce a manageable negotiating text. Officials therefore faced a mass of cumbersome negotiating text at the start of COP 6 (part I), which contributed to the slow progress in the negotiations. Indeed, many commentators pointed to the volume of text as one of the triggers for the failure of the session (e.g. Ott, 2001; Yamin and Grubb, 2001).

This points to the role played by texts not only as registers and reflectors of substantive progress, but as active tools for promoting such progress. Effective negotiation requires a *manageable* negotiating text, that is, one where the agreed and outstanding issues, along with the options open to parties, are clearly signposted and are not lost amid a sea of brackets and minutiae.

Chair's texts

The preparation of a proposed compromise text by the presiding officer – a Chair's or President's text – can mark a watershed in a negotiation process, often pointing the way to success or failure. It is the point at which a presiding officer exercises his/her authority to present his/her vision of a text that can spur parties to consensus. How far a presiding officer is prepared to go – creating an entirely new text or drawing faithfully on existing proposals, presenting a comprehensive compromise option or retaining alternatives on key points – is a matter of judgment. As the text is intended to move delegations closer to agreement, it must strike a delicate political balance between taking risks to propel the negotiations forwards, and not going so far as to impose an unwanted solution that is then rejected. If successful, the presentation of a Chair's text will mark the intensification of bargaining, or even the start of serious deal making.

A very important function of a Chair's text can simply be to produce a shorter, less complex, more manageable text, which clears out superfluous square brackets and unviable options, along with instances of duplication or editorial incoherence. The simple act of producing a document drafted in a single editorial style, using consistent terminology and language, can help negotiations to proceed in a more efficient manner, focused on the substance of the key political issues. An interviewee expressed the need for such an exercise as follows:

> There has to come a moment when the Chairman ... effectively says, 'we can't proceed interminably with ... all the options...I have to exercise my Chairman's prerogative to table a text that I believe properly and fairly reflects everybody's views, but cutting out a lot of the deadwood'... Most people in their heart of hearts appreciate it. However they posture, they basically appreciate it.

Another key function of a Chair's text can be to act as a *face-saving device*. As one interviewee noted, 'by that stage no party wants to withdraw something from the table so it's up to the Chairman to present...the face-saving compromise and choices'. By presenting text as his/her own, the presiding officer can help to loosen the bonds of ownership that tie parties to their own proposals. While parties might not be willing to back down in favour of the proposals of others, they can do so implicitly by supporting a Chair's text. Expressing support for a Chair's proposal is generally viewed as a noble gesture in the spirit of compromise, rather than as a capitulation (see also Chapter 4).

A very important factor in the successful production of a Chair's text is the timing of its release – too early, and it can be viewed as premature interference, too late, and there may be insufficient time to negotiate on the basis of it, and it may be seen as an unwelcome take it or leave it imposition. Another key factor is the extent of consultation undertaken by the presiding officer (or the organizers more broadly) with parties and coalitions in preparing the text. The correlation is not entirely straightforward. Extensive consultation may be helpful in securing legitimacy and acceptability for the text. On the other hand, it can create unreasonable expectations on the part of the delegates that their views will be reflected in the text, or otherwise erode the independence of the presiding officer's proposals. Before preparing a Chair's text, presiding officers will typically seek a mandate to do so from the negotiating body. As with all documents, the presiding officer will often receive strong support from the secretariat in preparing the Chair's text.

Chair's texts played a pivotal role in both the Kyoto and post-Kyoto negotiations. During the Kyoto negotiations, Chair Estrada brought out his Chair's text for AGBM 8 in late 1997, the last negotiating session before Kyoto. During COP 3 itself, he also made use of his authority as Chair to propose text on certain outstanding clauses.

In the post-Kyoto process, Chair's texts – more accurately, President's texts – were central to negotiations on a political deal covering the *package* of issues, which eventually led to the Bonn Agreements (see Table 11.4). COP 6 President Pronk produced his first President's text on the eve of the scheduled closure – and eventual failure – of COP 6. His second President's text was issued in the inter-sessional period before the resumption of negotiations at COP 6 (part II). This text included both an overview of the package of key political questions, and detailed compromise text on all aspects of the individual issues. President Pronk's third, and final, President's text on the package was released as ministers arrived for COP 6 (part II).

The negotiations on the *individual issues* of the flexibility mechanisms, compliance and LULUCF (and others) each saw differing levels of input by their presiding officers throughout the textual development process. On all three issues, however, the presiding officers prepared a Chair's text to go into COP 6 on the basis of inter-sessional consultations (see Tables 11.5, 11.6 and 11.7). All these Chair's texts, however, essentially comprised more or less extensive tidying up exercises, that is, aimed at making the text more manageable, rather than presenting compromises on difficult issues. This is not surprising, given that the post-Kyoto negotiations were fundamentally negotiations on a package deal, and no compromise Chair's text could have been prepared on one issue without reference to the others. These Chair's texts on individual issues were therefore very different to the Chair's texts that were produced either for the Kyoto Protocol negotiations, or for the post-Kyoto negotiations on the package deal.

The Chair's texts produced during the Kyoto and post-Kyoto negotiations enjoyed differing degrees of success. Although it was not without criticism, the Chair's text for the Kyoto Protocol negotiations (see Table 11.3) was widely welcomed as a vital contribution to moving the negotiations forward. One interviewee even suggested 'the Chairman's text was really the reason we got a Protocol'. Estrada's subsequent proposed compromise text on certain individual clauses at COP 3 itself were similarly absolutely critical in bringing parties to agreement.

The fortunes of the President's texts in the post-Kyoto negotiations differed considerably (see Table 11.4). The first President's text – the 'Note by the President' issued on the penultimate day of COP 6 – in effect precipitated the failure of the session. This failure cannot, of course, be attributed solely to a document, but many commentators have stated that the unorthodox approach taken in the preparation, presentation and subsequent handling of the Note contributed to the collapse of the negotiations (see Grubb and Yamin, 2001; Ott, 2001; Jacoby and Reiner, 2001; ENB, 2000b). The second President's text – the 'Consolidated Negotiating Text by the President' prepared for COP 6 (part II) – proved more useful, partly because the President's unusual approach to textual development had now started to sink in. However, this text – including both the overview and compromise text on all the issues – was only ever used as a 'tool', and never as the actual basis for negotiations. This role fell to the third President's text – the 'Core elements for the implementation of the Buenos Aires Plan of Action' prepared at the end of the first week of negotiations at COP 6 (part II). This proposed compromise deal on the package of key political post-Kyoto issues formed the basis for the Bonn Agreements adopted at that session.

The experience of Chair's texts in the Kyoto and post-Kyoto negotiations provide several valuable lessons, as discussed below.

How far to go?

The main issue when preparing a Chair's text is how far to go: whether to present a comprehensive compromise at one end of the spectrum, or a simple tidying up exercise at the other end. The climate change negotiations have seen various approaches in this respect. At one end of the spectrum, the Chair's texts on individual issues in the post-Kyoto negotiations consisted mostly of (very useful) tidying up exercises, with some language proposed by the Chairs on more technical issues. At the other end of the spectrum, the President's texts on the overall package deal of the post-Kyoto negotiations, as discussed below, sought to present a comprehensive compromise on all key political issues.

The Chair's text in the Kyoto Protocol negotiations fell somewhere in the middle of this range. For some, its chief achievement was simply to prune previous texts down to a workable size. The G-77 Chair, for example, in his opening statement to AGBM 8, expressed admiration that Estrada had 'been able to reduce a 200-plus-page compilation of proposals shrouded in a maze of brackets into a 25 page protocol' (Tanzania, 1997). In doing so, however, Chair Estrada also took a step further in producing *clean text*, that is, without alternatives, on the less controversial issues, notably on reporting, review, introductory and final clauses, and institutions and mechanisms. Estrada surmised that, on these articles, opposing views were not strongly held, and parties would accept the presentation of just one option, even if it were not their preferred one. In some cases, this helped directly to forge agreement. An example of this can be found in the provisions on amending the Protocol, where Estrada proposed to apply the same three-quarters voting majority as in the Convention in his Chair's text, removing the option of a two-thirds majority that had been supported by AOSIS. AOSIS did not insist, however, on reinserting their two-thirds voting majority option, and the issue was thus brought discreetly to closure. Clearing up less controversial issues in this way was important to ensuring that the final bargaining and deal-making stages would have the space and time to focus on the most difficult questions.

However, although Chair Estrada brought some of the more minor points to closure, he kept the major issues open. What he explicitly did not seek to do was to present a comprehensive compromise proposal. Instead, he retained several square brackets in the Chair's text as indicators of the core issues under negotiation, such as the level of the quantified target, and whether to adopt differentiated or uniform commitments. By maintaining these options, Estrada sought to ensure that he would not be accused of going too far, and would not lose the support of the different sides in the negotiations. On some controversial issues where alternatives were not so clear cut (e.g. joint implementation, general commitments for all parties, emissions trading), Estrada did seek to present a clean text which he thought might serve as a compromise. The success of this endeavour, however, was limited, partly because it was still too early for parties to consider such compromises.

One of the important aspects of the Chair's text was also that it derived almost all of its elements from previous negotiating texts. Delegates could broadly recognize its structure and contents and, because these were familiar, they aroused less suspicion. The draft article that did provoke controversy was precisely the one that departed significantly from existing proposals, namely that on voluntary commitments. While the subject matter itself was controversial, it can be argued that, if a text more in line with existing proposals had been included in the Chair's text, it would not have incited either such concern on the part of developing countries or so much critical attention from Annex I parties (see FCCC/TP/2000/2), and might have successfully found its way into the final Protocol.

A contrast to this approach in the post-Kyoto negotiations was the Note by the President presented on the penultimate day of COP 6. Here, President Pronk explicitly sought to present a comprehensive compromise, in his words, a 'balanced package', on all issues under the post-Kyoto negotiations. No options or alternatives were included, only the proffered deal. Needless to say, this was a risky strategy, given the contentious nature of the issues and the polarization of party positions. The result was that the paper was almost universally rejected for giving too much away to the other side – possibly an indicator that it did, in fact, strike the right balance. Nonetheless, it was the approach of putting forward a complete compromise that most parties were unhappy with, and indeed wary of, seeing it as a take it or leave it package.

Perhaps the most striking feature of the Note, however, was its unusual format and style, using narrative, short hand bullet points, rather than legal text. As Ott (2001, p281) put it:

> the text was not written in legal language, but contained political positions. Even if ministers at the conference had been able to agree on those positions, another meeting would have been necessary in order to translate these positions into treaty language.

The Note was thus virtually unrecognizable from the existing negotiating texts. Its issuance as a non-paper, rather than an official conference document, also confused delegates. As Grubb and Yamin (2001, p268) put it, 'It was an approach familiar in European Council meetings, but baffling to those used to UN negotiations; some found the jettisoning of...all the text they had developed and were familiar with – confusing and startling'.

Moreover, the Note included several very interesting, but novel and unprecedented, proposals, which had never before been tabled or discussed, and alone would have required days of negotiations. These two significant problems – total abandonment of existing texts and introduction of major new ideas at the last moment – meant that the presentation of a comprehensive compromise was unlikely to succeed. President Pronk adopted a different tactic for his two subsequent President's texts for COP 6 (part II), basing these much more closely on texts worked on in the negotiations themselves. The Core Elements paper, in particular, was based very closely on texts produced in the negotiating groups, so that, while it too presented a comprehensive 'deal', it was accepted as a basis for compromise.

Getting the timing right

Comparing the experience of the Kyoto and post-Kyoto negotiations also highlights the importance of timing in the production of Chair's texts. In the case of the Kyoto Protocol negotiations, Chair Estrada presented his Chair's text one full negotiating session before COP 3. This was sufficiently late to have allowed ample opportunity for parties to streamline and consolidate their own textual proposals as far as they could, but also gave parties time to familiarize themselves with the Chair's text, work with it, and make it their own before the final round of negotiations at COP 3. As one interviewee put it, 'people had a meeting to feel comfortable with the text, they thought about it in the inter-sessional period, and then went into Kyoto to have results'. Another commented: '[it wasn't easy to] manage Estrada's text with all its complexities...After a while, though...we familiarized ourselves with its structure and things became a lot clearer.'

In contrast, the first President's text for the post-Kyoto negotiations was circulated only on the evening of the second to last day of negotiations, less than 24 hours before the scheduled end of the Conference. Even assuming an extension of the deadline to Saturday lunchtime, there were only 40 or so hours physically available for negotiators to work on it. This was simply insufficient time to broker a complex deal, especially on such an unprecedented type of document. Because the President's Note came out so late, President Pronk presumably felt obliged to present a completely clean text, rather than one that retained a few options. However, precisely because it was so late, delegates saw the compromise text as an imposition upon which there was no time to improve.

The process of preparation

The process of preparing a Chair's text is also very important. The Chair's text in the Kyoto Protocol negotiations was prepared after considerable consultation, including an Expanded Bureau meeting convened specifically to review a draft of the text, private consultations between Chair Estrada and other trusted delegates (including on the margins a meeting of organized by Japan in Tokyo), and extensive discussions between Estrada and the secretariat. Indeed, the secretariat played a major role in preparing the text, producing the first draft of all but a handful of articles, and working closely with Estrada to prepare the final version. The Expanded Bureau meeting was of particular importance, helping to forge a sense of ownership of the text among those parties that would be most likely to challenge it so that they would be more inclined to look favourably upon the final version. Moreover, some changes were proposed at the meeting that helped to move the negotiations forward. For example, Japan and Samoa let it be known that they were dropping their demand for a carbon dioxide-only target, thus requiring just the options for a three and six gas target to be covered in the text.

Once the Chair's text had been drafted, Chair Estrada provided a briefing to each major coalition and NGOs immediately before AGBM 8 (the other subsidiary bodies were meeting at that time), explaining and justifying the approach to, and contents of, the text, and hearing any concerns raised by delegates. This was key to harnessing support for it. Not only did his briefing help delegates to understand better the motivations behind the structure and contents of the text, it also generated goodwill, particularly among less powerful parties, who were appreciative of his efforts. Zimbabwe, for example, speaking for the African Group at the opening of AGBM 8, commended Estrada 'for [his] efforts, and the time...put in informal consultations with many parties...in trying to explain the rationale behind the negotiating text' (AGBM 8, 1997c).

Preparation of the Chair's texts on the individual issues in the post-Kyoto negotiations, notably the Chair's texts prepared for COP 6, followed a similarly consultative approach, as they were drafted pursuant to inter-sessional informal consultations.

The preparation of the Note by the President at COP 6 was very different. In this case, the Note was prepared almost entirely by the COP President's Dutch team; there were no rounds of informal consultations or, as far as can be ascertained, extensive bilateral meetings, except with the ministerial Co-Chairs of the negotiating groups, and maybe a handful of trusted colleagues. Even the secretariat was not closely involved in the preparation of its substance. President Pronk presumably decided to keep his text specifically under wraps as it was being prepared, and *not* to consult on it. This meant it enjoyed no ownership when it came out – indeed, it had to contend with the considerable surprise of delegates at its approach – and faced a greater chance of rejection. Perhaps learning from this experience, the next President's text issued by President Pronk – the Consolidated Negotiating Text for COP 6 (part II) – was circulated and discussed in advance of the session at an open-ended meeting attended by over 350 delegates (see Box 9.1). This no doubt contributed to the more favourable reception of the text at COP 6 (part II) itself.

Working with the Chair's text

Once a Chair's text is published, the next important step is what to do with it. This is, of course, closely related to timing. In the case of the Kyoto Protocol negotiations, parties had time to work with the Chair's text at AGBM 8, with Chair Estrada urging delegates not simply to add square brackets and old language back into the text. His approach was largely successful. Brackets or footnotes recording objections and additional text were inserted, but in no case was text simply lifted out of previous negotiating texts and added straight back in. The outcome of AGBM 8 was therefore a larger text, some six pages longer, but it was a considerable achievement that it did not swell to a much greater size. As one interviewee put it, 'if...you throw out 150 points, people inevitably, when you bring it to the plenary...will say you missed this point or that...but even then if you just build in five points, you've got a bonus of 145'. The revised text that emerged from AGBM 8 was forwarded for final negotiation at COP 3. Estrada therefore allowed parties themselves to prepare their *own text* for the final negotiations at COP 3, but based on a more manageable foundation. He explained this approach in his final statement to AGBM 8 as follows:

> I did try to bring to you a proposal [the Chair's text] that was something in the middle of the road, something that will facilitate different positions to compromise, something that will help the understanding of different parties. And I'm satisfied with that; that was my task. The paper produced here [the revised text] has a different approach. It is the approach of people negotiating, trying to preserve positions on one side or another. I don't think this is bad, actually *it is part of the process, it is an instrument we need to progress, in the same way we need the instrument in between* (AGBM 8, 1997e, emphasis added).

Again, the contrast with the treatment of the Note by the President is a stark one. In this case, having circulated his President's text, President Pronk invited parties to submit written amendments on it. This encouraged parties to simply spend time identifying flaws in the document that they wanted to amend, rather than negotiating on it. As Grubb and Yamin (2001, p268) stated:

> no one knew how to comment on a presidential compromise text in a few hours without resorting to national positions. The net result was that the main groups spent the final official day of negotiations preparing written amendments to try to drag the new text back to their preferred positions – the exact opposite of what they should have been doing by this stage.

In commenting on the Note, parties showed a similar lack of self-restraint as they did in the negotiations on the flexibility mechanisms; the volume of written comments was massive, and it is essentially under these that COP 6 drowned.

The final negotiating texts

In every negotiation, there comes a crunch point when it is clear to delegates that the endpoint has been reached, and a deal must be struck (or abandoned). The negotiating text that the deal-making arena (see Chapter 9) has before it at that point is crucial. Most importantly, the text should faithfully codify all the deals and agreements reached to date, presenting familiar, almost entirely clean text, and only a small number of square brackets or alternative paragraphs denoting key outstanding issues. The aim is to focus the minds of negotiators (often ministers) solely on these issues, whose resolution will lead to the final compromise. A good example is the final deal-making round of the post-Kyoto negotiations on compliance at COP 6 (part II). Here, agreement had been reached on the President's text – the Core Elements paper – presented by the COP 6 President, except for the section on compliance. After extensive shuttle diplomacy (see Chapter 9), disagreement was whittled down to just two points. These were codified into a final draft text on compliance, composed of just a single page and two sets of brackets, issued as a non-paper. This provided a simple and effective tool for negotiators, who could put all other issues out of their minds. Agreement was reached in a 'friends' group based on that text.

A similar case occurred at COP 3, where the negotiating text going into the final meeting of the Committee of the Whole on the last day of the Conference was clean, except for two alternatives in the draft article on general commitments for all parties and the absence of any numbers in the annex listing targets. Although not all the remainder of the text had been agreed, the fact that it was issued as a clean document made it politically more difficult for parties to object to any part of it, unless they had very serious objections. This greatly simplified the final round of negotiations.

The contrast with COP 6 (part I) is clear. Here, the President's friends group, serving as a deal-making arena, was faced with a President's text – the

Note by the President – as a final negotiating text. The fact that its style and approach, along with much of its contents, represented a departure from previous negotiating texts, and that almost none of it had therefore been agreed, meant that it could not serve as an effective tool for facilitating agreement.

An important feature of final negotiating texts is that they are almost always issued in English only, due to lack of time for translation. This inevitably places non-Anglophone parties at a disadvantage. Although they are not being asked to formally adopt a decision – that will occur in plenary with a translated document – they are being asked to strike a deal, which it will then be extremely impolitic to renege upon. Of course, by this stage in the negotiations, the obstacles faced by non-Anglophones are more fundamental; even if final negotiating texts were translated, the negotiations would still be conducted exclusively in English (see Chapter 7). Although this state of affairs is generally tolerated by non-Anglophone parties, it can prove problematic. One of the concerns of the Russian Federation that delayed formal adoption of the Bonn Agreements at COP 6 (part II) was that the final deal on compliance had been struck based on a document only in English (see Chapter 9).

A similar issue arises with the status of documents used as final negotiating texts. In the Kyoto Protocol negotiations, the final text considered in the committee of the whole on the last night of COP 3 was issued as a CRP. This meant that, although informal, the document did enjoy official status, being made widely available, including on the internet, and kept in the secretariat and UN archives. However, the post-Kyoto negotiations have seen a tendency towards the greater use of non-papers as final negotiating texts. The COP 6 Note by the President was issued as a non-paper, as was the Core Elements paper at COP 6 (part II) and the final compliance text. With the exception of the COP 6 Note by the President, which was later appended to the COP 6 report, the other documents were not made widely available, and are no longer easily accessible. The problem is a wider one relating to non-papers, which are used extensively in the intensive bargaining and deal-making stages of negotiations. This raises important issues of transparency, as it is difficult for outsiders, or indeed future analysts, to scrutinize the final steps taken towards reaching a deal. It can also have implications for any future interpretation of decisions taken, as the textual history will be incomplete. Historical analysis of the negotiations on the CDM, for example, one of the most far-reaching articles in the Kyoto Protocol, is almost impossible, as its negotiations took place virtually entirely behind closed doors using non-papers.

Summary and concluding remarks

A key underlying message of this chapter is the importance of texts as a tool for the negotiations. Texts do not only reflect the status of negotiations, but can also help move the process forwards. They can do this in two main ways. Firstly, by codifying progress so that it becomes real. Through negotiations based on texts, agreement is built up and solidified as language is painstakingly approved word by word, comma by comma. Secondly, texts can serve as an indispensable tool to actually facilitate the work of negotiators. Given the complexity of the climate change negotiations, there is a great need for clear and manageable negotiating texts that are able to capture the state of play and highlight outstanding issues and options, thereby helping negotiators to focus their minds on the key questions requiring their attention. Otherwise, negotiators can often waste time simply trying to make sense of the text and the ideas expressed therein.

A carefully managed textual development process can also be critical in countering the inherent tendency of negotiators to procrastinate (see Chapter 12), encouraging delegates to progress from exploration to bargaining to deal-making. A key factor in the process is the balance the presiding officer and secretariat have to find between caution to maintain the acceptability of texts, and boldness to advance the negotiations. It is very important that the textual development process keep moving forward, and for each negotiating text to represent an appreciable advance on the last.

The chapter also underlines the importance of a judicious Chair's text in intensifying bargaining and launching deal-making. While a well-judged and well-timed document can help propel negotiations to agreement, a text that is ill-judged, late, or indeed too early, can literally bring negotiations to a halt. Like other elements discussed in this book, it seems that the key is finding a balance between prudence and innovation – putting forward proposals that help move negotiations on to a new stage, yet respecting the bounds of what is known, and familiar, and can be worked with.

Table 11.3 *Textual development of the Kyoto Protocol negotiations*

Type of document/title and symbol	Comment
Raw material	
Written proposals from parties	Over 430 pages of MISC documents
Precursor texts	
Synthesis of proposals by parties (FCCC/AGBM/1996/10)	Prepared for AGBM 5 in late 1996. Provides a narrative synthesis of proposals, indicating proponents.
Framework compilation of proposals by parties for the elements of a protocol or another legal instrument (FCCC/AGBM/1997/2 and Add.1)	Prepared for AGBM 6 in early 1997 Reproduces proposals verbatim under common headings, indicating proponents
Negotiating texts	
Report of the AGBM on the work of its sixth session: Proposals for a protocol or another legal instrument (FCCC/AGBM/1997/3/Add.1)	Prepared for AGBM 7 in mid-1997. Includes mostly legal language, and no longer indicates proponents. Prepared by consolidating the framework compilation at AGBM 6. 129 pages.
Report by the Chairs of the informal consultations conducted at AGBM 7 (FCCC/AGBM/1997/INF.1)	Outcome of work at AGBM 7. Issued as an INF (therefore with lower status) to encourage parties to focus on the forthcoming Chair's text. 82 pages
Chair's text	
Consolidated negotiating text by the Chairman (FCCC/AGBM/1997/7)	Prepared for AGBM 8 in late 1997. Sets out compromise language proposed by Chair Estrada on more technical points. Retains options on key political issues. 26 pages
Further final negotiating texts	
Revised text under negotiation (FCCC/CP/1997/2)	Outcome of work at AGBM 8 based on the Chair's text. Forwarded to COP 3. 32 pages
Non-paper by the Chairman of the Committee of the Whole (FCCC/CP/1997/CRP.2)	Issued at the end of the first week of COP 3 to inform ministers of the status of negotiations. 29 pages
Untitled (FCCC/CP/1997/CRP.4)	Issued on the penultimate day of COP 3. Included a first draft of individual emission targets. 24 pages
Final draft by the Chairman of the Committee of the Whole (FCCC/CP/1997/CRP.6)	Final deal-making text issued on the last day of COP 3. A virtually clean text, except for a handful of key political issues. 24 pages

Kyoto Protocol to the United Nations Framework Convention on Climate Change (decision 1/CP.3). 25 pages

Table 11.4 *Textual development on the package of key political issues in the post-Kyoto negotiations, resulting in the Bonn Agreements*[1]

Type of document/title and symbol	Comment
Raw material	
None	
Precursor texts	
Non-paper from the President-designate of COP 6 (FCCC/Non-paper, 2000a)	Prepared for the President's informal high-level consultations (Muiden, October 2000). First attempt at bringing all the key political issues on the post-Kyoto agenda together in one text
Non-paper from the President of COP 6 (FCCC/Non-paper, 2000b)	Prepared for the informal high-level consultations held on the middle Sunday of COP 6. Later circulated to all parties. Summarizes state of play
Note by the Co-Chairmen of the negotiating groups (FCCC/CP/2001/CRP.8)	Prepared during COP 6 (part II). Summarizes the state of negotiations, listing the main choices on the key political issues
Negotiating texts/chair's texts[2] Note by the President of the COP (Initially issued as a non-paper (FCCC/Non-paper, 2000c) then annexed to decision 1/CP.6)	Released on the penultimate day of COP 6 (part I). Sets out in conceptual terms President Pronk's proposed compromise on all key issues on the post-Kyoto agenda. Unusual approach and style.
Consolidated negotiating text prepared by the President (FCCC/CP/2001/2 and Add.1–6)	Prepared for the start of COP 6 (part II) by President Pronk. Presents a proposed deal on the key political issues, plus full compromise texts in legal language on all issues on the post-Kyoto agenda. Used as a tool for negotiations at COP 6 (part II)
Core elements for the implementation of the Buenos Aires Plan of Action (issued as non-paper)	Prepared during COP 6 (part II). Comprises a proposed compromise by President Pronk in legal language on the key political issues on the post-Kyoto agenda. Formed basis for final deal-making
The Bonn Agreements on the implementation of the Buenos Aires Plan of Action (decision 5/CP.6). 15 pages	

Note: The texts listed in this table are in addition to the separate texts issued on individual issues in the post-Kyoto negotiations. The textual development of the flexibility mechanisms, compliance and LULUCF is set out in Tables 11.5 to 11.7.

Note: 1 All negotiating texts were issued as Chair's – in this case President's – texts.

Table 11.5 *Textual development on the flexibility mechanisms in the post-Kyoto negotiations, resulting in the Marrakesh Accords*

Type of document/title and symbol	Comment
Raw material	
Written proposals from parties	Over 740 pages of MISC documents (from COP 4)
Precursor texts	
Synthesis of proposals from Parties (FCCC/SB/1999/INF.2 and Add.1–3)	Prepared for SBSTA/SBI 10 in mid-1999. Collates proposals together under common headings, indicating proponents. English only, except for glossary of terms in all UN languages
(Revised and consolidated) Synthesis of proposals from Parties (FCCC/SB/1999/8 and Add.1)	Prepared for SBSTA/SBI 11 in late 1999. Revises the earlier text, including new proposals, square brackets and mostly legal language
Negotiating texts Text for further negotiation (FCCC/SB/2000/3)	Prepared for SBSTA/SBI 12 in mid 2000. Goes further in merging proposals and using square brackets. Proponents still indicated. 134 pages
Consolidated text (FCCC/SB/2000/4)	Prepared for SBSTA/SBI 13 (part I). Outcome of negotiations at SBSTA/SBI 12, including deletion/merging of proposals, and some text proposed by the Chairs. Proponents not indicated. 118 pages
Consolidated text (FCCC/SBSTA/2000/10/Add.1)	Outcome of SBSTA/SBI 13 (part I). 162 pages
Chair's texts Text by the Chairs (FCCC/SB/2000/10, Add.1–4)	Prepared for COP 6, based on previous text and inter-sessional consultations. Further merging/consolidation of proposals by Chairs. 83 pages
Consolidated negotiating text prepared by the President (section on flexibility mechanisms) (FCCC/CP/2001/2/Add.2) (see Table 11.4)	Proposed compromise text on all aspects of the flexibility mechanisms prepared by President Pronk for COP 6 (part II). Used as a tool at that session. 38 pages
Further and final negotiating texts Texts forwarded by the subsidiary bodies to COP 6 (part I). (FCCC/CP/2000/INF.3 (Vol V))	Outcome of first week of COP 6 (part I). 84 pages
Texts forwarded to the resumed 6th session by COP 6 (part I). (FCCC/CP/2000/5/Add.3 (Vol V))	Outcome of COP 6 (part I). 95 pages
Draft decisions on which progress was noted by COP 6 (part II) and which the COP decided to forward to its 7th session for elaboration, completion and adoption. (FCCC/CP/2001/5/Add.2)	Outcome of COP 6 (part II), forwarded for final negotiation at COP 7. 46 pages

Five decisions in final Marrakesh Accords (Decisions 15/CP.7–19/CP.7). 70 pages

Table 11.6 *Textual development on compliance in the post-Kyoto negotiations, resulting in the Marrakesh Accords*

Type of document/title and symbol	Comment
Raw material	
Written proposals from parties	Over 250 pages of MISC documents (from COP 4)
Precursor texts	
Elements of a compliance system and synthesis of submissions (FCCC/SB/1999/7 and Add.1)	Prepared for SBSTA/SBI 11 in late 1999. Presents a synthesis of submissions under common headings, and a listing of possible elements for a compliance system. Submissions mostly in response to questions issued at SBSTA/SBI 10. Proponents indicated
Note by the Co-Chairs of the JWG (FCCC/SB/2000/1)	Prepared for SBSTA/SBI 12 in mid 2000. Presents proposals under common headings, including a simple note where discussion is insufficiently advanced to identify options
Untitled text on compliance (Annex to SBI 12 report)	Outcome of work at SBSTA/SBI 12. Has characteristics of a negotiating text, but was not used at any session. Instead, it was 'further developed' by the Co-Chairs into a new text for SBSTA/SBI 13 (part I) (see below)
Negotiating texts	
Proposals from the Co-Chairs of the JWG (FCCC/SB/2000/7)	Prepared for SBSTA/SBI 13 (part I) based on the previous text. Includes legal language and square brackets. 26 pages
Proposals from the Co-Chairs of the JWG (FCCC/SBI/2000/10/Add.2)	Outcome of SBSTA/SBI 13 (part I). 20 pages
Chair's text	
Text proposed by the Co-Chairs of the JWG (FCCC/SB/2000/11)	Prepared pursuant to inter-sessional informal consultations for COP 6 (part I). 18 pages
Consolidated negotiating text prepared by the President (on compliance). (see Table 11.4) (FCCC/CP/2001/2/Add.6).	Proposed compromise text on all aspects of compliance prepared by President Pronk for COP 6 (part II). Used as a tool at that session. 14 pages
Further/final negotiating texts	
Texts forwarded to the resumed 6th session by COP 6 (FCCC/CP/2000/5/Add.3 (Vol IV)	Outcome of COP 6 (part I). 30 pages
Draft decisions on which progress was noted by COP 6 (part II) and which the COP decided to forward to its 7th session for elaboration, completion and adoption (FCCC/CP/2001/5/Add.2)	Outcome of COP 6 (part II), forwarded for final negotiation at COP 7. 11 pages

Decision in final Marrakesh Accords (decision 24/CP.7). 13 pages

Table 11.7 *Textual development on LULUCF in the post-Kyoto negotiations, resulting in the Marrakesh Accords*

Type of document/title and symbol	Comment
Raw material	
Written proposals from parties	Over 680 pages of MISC documents (from COP 4)
Precursor texts	
List of policy and procedural issues related to Article 3.3 and 3.4 (FCCC/SBSTA/1999/5)	Prepared for SBSTA/SBI 10. Narrative overview of the issues based on submissions
Consolidated synthesis of proposals made by parties (FCCC/SBSTA/2000/9)	Prepared for SBSTA/SBI 13 (part I), following request at SBSTA 11. Compilation of party submissions under common headings. Proponents indicated. No consolidation. No square brackets
Negotiating text	
Recommendation by the SBSTA (FCCC/SBSTA/2000/10/Add.2)	Outcome of SBSTA/SBI 13 (part I). Legal language. Square brackets. 14 pages
Chair's texts	
Text by the Chair (FCCC/SBSTA/2000/12)	Prepared pursuant to inter-sessional informal consultations for COP 6 (part I). 12 pages
Consolidated negotiating text prepared by the President (section on LULUCF). (see Table 11.4) (FCCC/CP/2001/2/Add.3/Rev.1)	Proposed compromise text on all aspects of LULUCF prepared by President Pronk for COP 6 (part II). Used as a tool at that session. 6 pages
Further/final negotiating texts	
Texts forwarded to the resumed 6th session by COP 6 (FCCC/CP/2000/5/Add.3 (Vol IV)	Outcome of COP 6 (part I). 11 pages
Draft decisions on which progress was noted by COP 6 (part II) and which the COP decided to forward to its 7th session for elaboration, completion and adoption (FCCC/CP/2001/5/Add.2)	Outcome of COP 6 (part II), forwarded for final negotiation at COP 7. 9 pages

Decisions in final Marrakesh Accords (11/CP.7) 9 pages

Time Management

Chairman Estrada... after stressing over and over again the extraordinary amount of business to be got through in such a short time, promptly adjourned the session after little more than an hour (ECO, 1995).

Introduction

The efficient management of time is a critical dimension to the organization of any negotiation process, and especially so for highly complex global negotiations. While the amount of time available is, of course, a consideration, how that time is used is equally, if not more, important. The key task in any time management exercise is to counter the natural tendency of delegates to backload negotiations – to delay the start of deal-making, or even intensive bargaining, until the last possible moment – while at the same time meeting their need to engage in thorough exploration of the issues in the initial stages. The Kyoto Protocol and post-Kyoto negotiations provide us with two contrasting and instructive cases in this regard: the Kyoto negotiations ended with agreement within the deadline at COP 3 in 1997 at the very last minute, while the post-Kyoto negotiations basically ran out of time at the scheduled COP 6 finale, having to go through two further finales before finally striking a deal at COP 7 in 2001 a year later. Even then, it was only at COP 9 in 2003 that the full slate of issues under the post-Kyoto negotiations was finally resolved. This chapter explores the issue of time management, comparing the experiences of the Kyoto and post-Kyoto rounds. It first discusses the amount of time available to both negotiating rounds, before considering how that time was used and how time management tools were wielded. The chapter concludes by discussing the implications of 'negotiation by exhaustion'.

Duration

The Kyoto Protocol negotiations took place over a period of 32 months, from the adoption of the Berlin Mandate in April 1995 to the adoption of the Kyoto Protocol in December 1997. The negotiating body, the AGBM, met eight times within this period, plus the half-day resumed session immediately before COP 3. With COP 3 itself, this made up a total of nine meetings. A review of other recent environmental negotiations (see Table 12.1) suggests that the duration of the Kyoto Protocol negotiations was neither particularly long nor short, falling roughly in the centre of the wide range of negotiating time spans. The intensity of the process, however, was notable, with a high ratio of meetings relative to the duration of negotiations.

Table 12.1 *Comparison of duration with other environmental negotiations*

Negotiation	*Duration in months*	*Meetings*
UNFCCC	17 (December 1990 to May 1992)	6
Kyoto Protocol	32 (April 1995 to December 1997)	9
Bonn Agreements and Marrakesh Accords	24 (Initial deadline) (November 1998 to November 2000)	5
	32 (Adoption of Bonn Agreements) (November 1998 to July 2001)	6
	36 (Adoption of Marrakesh Accords) (November 1998 to November 2001)	7
Stockholm Convention on Persistent Organic Pollutants	30 (June 1998 to December 2000)	5
Cartagena Biosafety Protocol	39 (Initial deadline) (November 1995 to February 1999)	6
	50 (Adoption) November 1995 to January 2000)	7
Framework Convention on Tobacco Control	33 (May 2000 to February 2003)	6

Although the Kyoto Protocol negotiations were concluded within the deadline, they did produce a considerable amount of unfinished business – including the details of the flexibility mechanisms, provisions on the LULUCF sector, the compliance system, and reporting and review methodologies – which then made up the post-Kyoto negotiating agenda. The existence of unfinished business at the close of a negotiation is not unusual and is often part and parcel of the continuous negotiation process at work in many environmental regimes. What is unusual in the case of the Kyoto Protocol, however, is that its unfinished business affected the effort required of parties to meet its emission targets to such an extent that most Annex I parties, the key players in the regime at this stage, delayed their ratification of the Protocol until a deal was reached in the post-Kyoto follow-up negotiations three-and-a-half years later.

It is unlikely, however, that more time would have enabled the resolution of (more of) the Protocol's unfinished business before the treaty's adoption. On the contrary, attempting to resolve more details through a longer process would have increased the complexity and opportunity for disagreement and obstruction, perhaps threatening the adoption of the Protocol itself. One interviewee, who began by arguing that a single extra meeting might have resolved certain questions, ended up querying:

but having said that…would it have come together? Would other things not have intervened? Could we have had other questions? This morass of details…Oh, my goodness, throw up your hands, we can't get there.

The exception concerns the specific issue of sinks (the LULUCF sector), where some interviewees argued that an earlier start to the negotiations on that issue, which did not really get going until AGBM 8 in late 1997, might have produced a better result.[1]

The Kyoto negotiations thus implicitly initiated a *two-stage process*. The political deal and basic structure of the Protocol were first agreed in Kyoto, followed by a second stage of negotiations on the implementation details of the Protocol in the post-Kyoto round. The more detailed work carried out as part of the post-Kyoto negotiations was arguably only possible based on the broad political agreements reached in Kyoto. As one interviewee commented:

A lot of the things we did resolve in Kyoto were political and I think that the details could only be done once you got the political decisions out of the way. So even if you had another two years to do it, you wouldn't have got any further forwards. You would have had to resolve the political decisions first.

The initial deadline for the subsequent post-Kyoto negotiations allowed for two years of negotiations – from COP 4 in November 1998 when the Buenos Aires Plan of Action was adopted, to COP 6 in October/November 2000 – with a total of five (initially only four – see below) negotiating sessions. The deadline of COP 6 and the small number of meetings was, with hindsight, excessively optimistic, and made for an extremely tight schedule for reaching agreement on the wide-ranging post-Kyoto agenda. It was set with (laudable) political intentions, that is, to enable countries to ratify the Protocol and for it to enter into force as soon as possible. A time span of two COPs seemed appropriate – the negotiations on the Kyoto Protocol had similarly been launched at COP 1, with a deadline of COP 3. However, the great technical work and, more importantly, evolution of understanding that was needed on the unprecedented issues raised by the Protocol meant that two years and four/five meetings was unrealistic; negotiation of the Kyoto Protocol itself had required more than double the planned number of negotiating sessions. The small number of negotiating sessions was compensated for to some extent by the proliferation of workshops but, as discussed in Chapter 10, these were not fully inclusive and could not actually engage in bargaining or take any decisions.

On the technically highly complex issue of LULUCF, for example, it was not until May 2000 that the IPCC produced its landmark Special Report on LULUCF (IPCC, 2000) requested at COP 4 (and that was a remarkably quick delivery), leaving only five months and two negotiating sessions to study the document before the COP 6 deadline. This meant that the bulk of proposals from parties on LULUCF were received in just the three months before COP 6 (see also Chapter 11). On the flexibility mechanisms, there was a real need for parties, especially developing countries, simply to understand the basic concepts that were being discussed and to gain ownership of those concepts before real negotiations could start. It is worth recalling that the negotiations on the flexibility mechanisms were attempting to create entirely new markets and economic systems, the likes of which can only be compared to the trading architecture under the WTO, which took decades to design. Where the deadline was least realistic was for the reporting, review and accounting provisions; clearly, reporting and review guidelines could only be finalized once the substance of commitments was known. Logically, a COP 6 deadline for provisions on LULUCF and the flexibility mechanisms, for example, should have been accompanied by a COP 7 deadline for reporting requirements on this sector. However, in an attempt to sustain political momentum after Kyoto, the timeline was simply not thought through when it was set at COP 4, and COP 6 became the universal deadline.

Interestingly, the actual date of COP 6 was not confirmed until COP 5. This was because certain parties wanted to delay the COP until early 2001, after the US Presidential election, at which point it was deemed that the US would be in a better position to take decisions. Other parties, however, notably the G-77 and China, preferred to adhere to the strict letter of the Convention, which requires an annual session (albeit 'unless otherwise decided' – Article 7.4). Deciding to hold COP 6 in 2000 was seen by many at the time as a signal of determination to forge ahead with the implementation of the Convention and the Protocol (ENB, 1999). However, as well as leading to uncertainty over the nature of the US administration, the 2000 deadline meant one less negotiating session.

Growing concern over lack of time led delegates to agree to convene what amounted to an extra negotiating session – SBSTA/SBI 13 (part I) in September 2000. This was made possible within the allotted budget due to extra money from donors, the offer from France to host the session and, very importantly, cutting one week of interpretation time from the scheduled session of SBSTA/SBI 12 in June 2000 and allocating it to SBSTA/SBI 13 (part I). This was an extremely innovative response to the need for more time: the first weeks of SBSTA/SBI 12 and SBSTA/SBI 13 (part I) were simply designated as a series of 'informal workshops', that is, conducted through informal meetings without interpretation, but in effect allowing the continuation of negotiations across an extra two weeks.

Famously, however, agreement was not reached at COP 6. One more negotiating session – COP 6 (part II) – was needed to strike a deal on the Bonn Agreements, and a second – COP 7 – was needed to reach agreement on the more technical details of the Marrakesh Accords. The political dimension of the supposedly more 'technical' post-Kyoto negotiations became almost as acute as the supposedly more 'political' negotiations on the Kyoto Protocol itself that had preceded them. This suggests that there is rarely any such thing as a purely 'technical' negotiation. The post-Kyoto negotiations were eventually resolved through their own two-stage process: first the political deal was struck at COP 6 (part II) with the Bonn Agreements, allowing, second, the more technical aspects of the negotiations to conclude based on that deal with the Marrakesh Accords. This was not the intent at the start of the post-Kyoto negotiations, but rather a deliberate strategy by COP 6 President Pronk to sequence the finale of the negotiations in this way, once he surmised there would be insufficient time to resolve the whole swathe of issues under negotiation. The deliberate separation of the political from the technical in this way proved problematic at COP 6, largely because it was an approach unusual to the climate change delegates. Nevertheless, once the approach had been understood, it proved more successful for COP 6 (part II). Sequencing decisions on (more) political and (more) technical questions over successive time periods – so key political issues are identified and resolved first, providing the basis for subsequent more technical discussions – can provide a means of managing the intricacies of very complex negotiations.

The use of time

The wide variation in the duration of negotiations shown in Table 12.1 suggests that, beyond a reasonable point, the amount of time available to a negotiation is less important than how that time is used. As Kaufmann puts it:

> it is not possible to establish an automatic and generally valid causal relation between the length of a conference and its degree of success or failure. What can be said is that the Chairman...and the delegates should have a clear idea how to organize their work in terms of the available time (Kaufmann, 1989, p52).

A key dimension to the effective use of time in any multilateral negotiation is to address the propensity for parties to backload negotiations, in other words, to hold off intensive bargaining and deal-making until the last possible moment. This is a deep-seated tendency among negotiators at all levels (e.g. see Kaufmann, 1989; Fisher et al, 1992; Illich, 1999). As one interviewee commented:

> things go very slowly when we still have a lot of time. When we approach a deadline, the discussion accelerates...and we finish in the last days at midnight...I don't like it, but I'm afraid it is so.

The backloading of negotiations is essentially a feature of brinkmanship; parties putting off making concessions as long as possible, in the hope that others will back down first. According to one interviewee, 'they [the negotiators] will trade in their chips at probably the latest possible time in the hope that it will maximize the benefit they can extract from it'. Negotiators, therefore, are highly reluctant to show much flexibility until the deadline is almost upon them.

In terms of time management, a negotiation can be crudely divided into three (see also Chapter 2 and Table 11.2): the first stage involving exploration of the issues, the second stage when proposals are tabled and low-key bargaining may get underway, and then the finale, where the bulk of bargaining takes place and deals are (hopefully) struck. Although attention often focuses on the finale, when negotiations are at their most intense, the careful management of time prior to that point is also critical, providing the conditions for a successful finale to unfold. According to Kaufmann, 'delegates must make careful use of the time available to prepare for the moment of truth, the last days of the conference when the principal decisions must be taken' (Kaufmann, 1989, p52).

The first and second stages

In most negotiations, the first couple of sessions are dedicated to exploration of the issues. The AGBM, for example, embarked on an explicit process of 'analysis and assessment' for a year between COP 1 and COP 2. The post-Kyoto LULUCF negotiations similarly began by inviting parties to present data and commissioning the IPCC to prepare a Special Report. The exploration stage can be long, often taking up half of the available meeting time. This can be frustrating, as little progress in forging an agreement is apparently registered. Reaching agreement in a global negotiation, however, requires not only substantive bargaining and deal-making, but also a preparatory learning process that lays the groundwork for that bargaining through investigation and study of the issues, often accompanied by considerable posturing on the part of delegates. As one interviewee noted with respect to the Kyoto Protocol negotiations, 'we had to go through two years of learning, figuring out other positions'. Time is required simply for issues to ripen in the minds of parties, even if little substantive analysis actually takes place in the negotiating forums. Moreover, there is a sense that delegates may need to 'talk themselves out' before bargaining can start. Chair Estrada explained in his interview: 'we opened the possibility [in the Kyoto Protocol negotiations] for people to repeat things one thousand times and get everybody tired, which is necessary...to start working'. Bodansky makes a similar comment on the Convention negotiations that is worth quoting in full:

> This sparring process, although frustrating to those seeking rapid progress, played a necessary role by giving states an opportunity to voice their views and concerns. They learned about and gauged the strength of other states' views. They sent up trial balloons and explored possible areas of compromise. Indeed, without this mutual learning process, it is hard to imagine that agreement would have been possible (Bodansky, 1993, p475).

The need for exploration and learning was particularly great for the post-Kyoto negotiations, which were faced with a set of novel issues with complex implications. Time for reflection, study and the exchange of ideas was therefore extremely important to bring all delegations roughly up to the same level simply in terms of understanding what had been agreed in the Kyoto Protocol, and what issues now needed to be resolved, let alone what the substantive options might be. The various inter-sessional informal workshops (see Chapter 10) that took place were very important in this respect. The need for exploration is also reflected in the large number of miscellaneous ('MISC') documents that were issued in 1998, especially on the flexibility mechanisms, representing an outpouring of views and opinions (see also Chapter 11).

The second stage of a negotiation is marked by the tabling of proposals and the onset of bargaining. The tabling of proposals typically takes place through the submission of text, which is then published in MISCs (see Chapter 11). Once proposals have been tabled, bargaining can commence. At first, this bargaining will be tentative, focused on the least controversial issues, or conducted through highly unofficial, informal channels. It will, however, gradually intensify, and may even lead to deal-making on more technical matters as the finale approaches. The most politically difficult questions are rarely the subject of serious bargaining at this stage, although initial, very informal discussions may take place behind the scenes among key players. During the Kyoto Protocol negotiations, for example, a small group of Annex I parties worked together on emissions trading behind the scenes of the latter AGBM sessions. In addition, secret contacts also began, at the informal consultations hosted by Japan, on what became the CDM.

Of course, the sequencing of stages is rarely clear cut, and stages will often overlap. Exploration of issues is likely to continue late on into the process as new problems or opportunities emerge, while some parties may delay tabling their proposals until the finale is very close and bargaining has actually started. In the Kyoto Protocol negotiations, for example, the US did not table its proposed emission target until the last session of the AGBM in late 1997. The late tabling of proposals is not a problem if promising new compromise ideas are injected into the talks, based on developments in the negotiations. It is detrimental, however, if entirely new concepts are put forward when there is insufficient time to consider them, or if the volume of late proposals is such that it complicates the negotiations and impedes the search for compromises. As the negotiations approach their finale, the focus of parties' efforts should be not on devising and tabling proposals, but on working with existing proposals to try to find integrative solutions.

The finale

The finale of a negotiating round will unfold at the last scheduled negotiating session before the deadline runs out. The finale of the Kyoto negotiations – COP 3 – lasted 10 days, while the scheduled post-Kyoto finale – COP 6 – lasted two weeks, as did the subsequent finales at COP 6 (part II) and COP 7. Because of the natural tendency of parties to backload the negotiation process, most of the bargaining and deal-making will take place only during the finale, however long the previous two stages have been. According to one interviewee, speaking about the Kyoto Protocol negotiations, 'actually before Kyoto nothing happened. The whole Protocol negotiation was just in those two weeks.'

Even if almost all of the intensive bargaining and deal-making is concentrated in the short period of the finale, the groundwork carried out during the first two stages is crucial to paving the way for a manageable and successful finale. When the finale starts, parties should have already made the transition from exploration and the tabling of proposals to bargaining, and be poised to engage in intensive bargaining and deal-making from the opening of the negotiating session. All the main proposals should be on the table, the broad range of options should be clear, some of the less contentious issues should have been preliminarily resolved, there should be a manageable negotiating text (see Chapter 11), and broad agreement should prevail on the structure of the negotiation process (see Chapter 9). The finale can then hit the ground running.

A common feature of finales is the way in which deal-making on the most important issues is typically backloaded until the *last night* of the negotiating session, often overrunning the deadline by a few hours. Indeed, the last night often constitutes a stage in the negotiation process in its own right. It is not surprising that this should be the case; in highly complex processes with many inter-linkages among issues it is impossible to agree on the final, core political questions except as a package, in quick succession. A marathon last night is often the end point of a punishing few days where bargaining and deal-making have intensified with the approach of the deadline, and talks have continued at a gruelling pace, often late into the night for days on end (Yamin and Depledge, 2004). This all-too-common dimension of the finale has been termed 'negotiation by exhaustion' (see Oberthür and Ott, 1999). It is discussed separately in more detail below.

Time management tools

A constant danger in a negotiation process is that it may get stuck in a particular stage – incessantly exploring issues, not making the transition to bargaining, avoiding deal-making – either due to the efforts of obstructionist parties or through simple procrastination and excessive brinkmanship. As one interviewee admitted, 'our one failing as negotiators is that...we sometimes think the process could go on forever'. It is thus vital for the organizers of the negotiation process – the presiding officers and secretariat – to place constant pressure on parties to move ahead in the negotiations, and to instil dynamism and momentum to offset the natural propensity to backload the negotiations.

The organizers can wield a number of tools to this end:

- use of rhetoric in verbal and written addresses, that is, pushing parties to advance the negotiation process
- exerting leadership to move negotiations up a gear in procedural and organizational terms, notably through the convening of new negotiating arenas and preparation of more advanced negotiating texts
- use of symbolic breaks and markers to signal an intensification of the negotiation process
- in all of the above, talking up, and appealing to, the deadline.

We now explore how time was used in the Kyoto Protocol and post-Kyoto negotiations, and in particular how the organizers wielded the time management tools mentioned above in these two very different negotiating rounds.

Time management in the Kyoto Protocol negotiations

Time management proved extremely important to the Kyoto Protocol negotiations. From the outset, Chair Estrada had a good idea of how the negotiations should proceed over time, and what particular steps needed to be taken at what time in order to meet the COP 3 deadline. His time management strategy was helped by early clarity over the deadline and the timing of COP 3, along with the fact that Estrada had overall responsibility for the whole negotiation process.

The first two stages

Chair Estrada used particularly strong rhetoric in his verbal addresses throughout the Kyoto Protocol negotiations to place pressure on parties to advance the negotiation process. Such pressure was often expressed by urging delegates to 'start negotiating', that is, bargaining, in contrast to 'repeating positions'. Estrada also put pressure on individual parties in this way, both through public statements and bilateral contacts. He regularly chided the US and Japan for their delay in proposing emission targets, and even wrote to the EU Presidency to urge the EU to provide prompt clarification on how its proposed bubble arrangement[2] would work. From 1997, Estrada made extensive use of negotiating arenas and negotiating texts to push parties to move on from exploration and initiate bargaining. For example, his exclusion of NGOs from the informal non-groups convened at AGBM 6 in early 1997, which many delegations and NGOs considered premature (see Chapter 14), sought to signal to parties that now was the time to start bargaining. Similarly, at AGBM 8, Estrada used late night meetings to raise the tempo of the negotiations; for one interviewee, 'late night meetings were important to cranking up the pressure'. The production of gradually more advanced texts – precursor texts, the negotiating text and then the Chair's text (see Chapter 11) – similarly pushed parties to intensify their work. The six-month rule, whereby the negotiating text for the Protocol had to be circulated at least six months before its adoption, was very important in this regard, providing a concrete, legal watershed – as well as a useful symbolic break – after which point bargaining could be expected to commence.

COP 2, which took place roughly midway in the AGBM negotiations, was built up by the organizers of the negotiation process into another very useful symbolic break. The organizers reinforced the natural break point that it provided by organizing a ministerial roundtable and promoting the Geneva Ministerial Declaration which, notwithstanding its shortcomings, acted as 'a further impetus' (FCCC, 2000) to the negotiations (see also Chapter 8). The Executive Secretary commented in his closing statement to COP 2 that the 'political content' of the session had 'exceeded his expectations' (ENB, 1996, p12). Indeed, although no decisions were taken at COP 2, the session marked a pivotal shift in the direction of the negotiations, as the US declared its commitment to adopting targets accompanied by emissions trading. This, combined with the progressive Geneva Ministerial Declaration, added up to a very important political statement of the commitment of the vast majority of parties to success in Kyoto. In addition, Chair Estrada and the secretariat billed AGBM 4 (taking place during the same period as COP 2) as a session for 'taking stock and intensifying efforts' (AGBM 4 report), that is, an opportunity to look back on the negotiations to date and then progress to a more intensive stage. This symbolic break was heavily engineered by the organizers, who succeeded in formalizing it into the AGBM 4 report with the statement 'the emphasis of the work of the AGBM must now move progressively towards negotiation' (AGBM 4 report, p10).

In all their efforts to instil a sense of urgency in parties, the organizers appealed relentlessly to the COP 3 deadline. The importance of deadlines in generating decisions is recognized in the literature. Zartman and Berman (1982, p195), for example, demonstrate how deadlines 'tend to facilitate agreement, lower expectations, call bluffs and produce final proposals'. A less positive effect of deadlines is to push parties to brinkmanship and 'negotiation by exhaustion', a problem discussed further below.

The COP 3 deadline for the Kyoto negotiations was, to a large extent, artificially generated and self-imposed. It did not, for example, coincide with a major intergovernmental conference, such as the Earth Summit in 1992, which had provided a politically important deadline for the negotiation of the Convention. Moreover, the wording of the Berlin Mandate was quite soft, stating only that the Protocol should be completed 'with a view' to its adoption at COP 3, rather than a stronger statement, such as 'shall be adopted by COP 3'. It was therefore critical to the deadline's authority that it should be continuously reinforced through the actions and words of the organizers of the negotiation process. Almost every AGBM report, for example, made reference to the COP 3 deadline. Moreover, neither Estrada nor the secretariat ever raised the possibility that the negotiations might fail. In the run-up to COP 3, the Japanese delegation privately approached the secretariat to explore possible contingency plans if the Conference failed, to which the secretariat responded 'we have not thought about contingencies. For us, it is the unthinkable' (secretariat e-mail, 1997). The Executive Secretary took a hard line on this. A suggestion by an AGBM team member to calculate the costs of reconvening COP 3 so as to demonstrate how expensive failure would be was rejected on the grounds that, if the possibility of a get-out clause were raised, the imperative of reaching agreement would dissipate.

The deadline of COP 3 thus became unquestioned, with the political necessity of fulfilling it one of the few common goals publicly shared by parties negotiating in good faith. The deadline generated its own momentum, pulling parties to agreement. Its political and public visibility meant that missing the deadline would have been viewed as a political disaster on the international stage, with OECD governments fearing that they would be held responsible at the national level. Interviewees agreed that the pressure of the deadline, and the operationalization of that pressure through constant reinforcement, had been critical to reaching agreement.

The overall impact of time management during the first and second stages of the Kyoto Protocol negotiation process was thus to help create the conditions that would enable the COP 3 finale to reach agreement. The negotiation process had not stood still throughout the 32-month process, but had moved steadily ahead, propelled largely by the actions of the organizers. By the time COP 3 started, parties had a manageable, agreed negotiating text with which to work, and an accepted structure for the organization of the negotiations; they had got to know one another and built up trust in the production team; they had gradually increased the intensity of their deliberations and identified the options among which they had to choose. Significantly, they had also started to engage in bargaining and clear up some of the less contentious issues. Negotiations based on the Chair's text at AGBM 8, for example, resulted in preliminary agreement on many of the legal clauses and institutional questions, along with the basic framework of reporting and review provisions. Even on more controversial questions, substantive bargaining based on the Chair's text allowed the options facing parties to be more clearly defined in the revised text that was forwarded to COP 3. In some cases, even if no preliminary agreement was reached before COP 3, much of the language in the revised text was eventually incorporated into the final Protocol (see FCCC/TP/2000/2).

The extent to which the organizers of the negotiation process were able to induce parties to accelerate their bargaining and deal-making should not, however, be overstated. Estrada was thwarted in his desire to secure a mandate to prepare a negotiating text for AGBM 6 (the mandate was eventually granted for AGBM 7), the US still bided its time to announce its proposed targets until AGBM 8, while bargaining on the key questions – the nature, level and timing of targets, flexibility mechanisms, developing country issues – was all backloaded to the negotiation finale of COP 3. Nevertheless, where the organizers could make a difference – encouraging the production of a manageable negotiating text, instilling a sense of finality with regard to the deadline – judicious time management enabled them to do so.

The finale

During the COP 3 finale, Chair Estrada continued to wield the time management tools that he had done in the initial stages of the negotiations, and added a few extra ones, reflecting the absolute imperative, at this final stage, for the process not to get stuck. Estrada continued to make extensive use of rhetoric, seeking to maintain a sense of urgency in the minds of delegates, and make sure they were aware of time ticking away. In doing so, he appealed repeatedly to the COP 3 deadline, emphasizing over and over again that agreement must – and would – be reached by that time. He also appealed to the arrival of ministers for the second week, urging parties to have resolved all but the most difficult political issues by that time. The high level segment, taking place in the last three days of the conference when negotiations are typically at their most intense, provided an excellent natural break point for parties to work towards, then take stock, and subsequently renew their efforts with additional political input from ministers.

Another important symbolic marker was the designation of a day of rest – the middle Sunday – when no official meetings at all took place. This was very important in providing an opportunity for all those involved in the negotiations –

including the organizers – to recharge their batteries for the coming difficult final days. It also provided the space for parties to brief their arriving ministers, take stock of the progress to date, consider their strategies for the coming days, and even consult informally with allies and adversaries. The rest day thus served as a useful launch pad to the start of more intensive negotiations during the high-level segment.

Throughout COP 3, Chair Estrada made use of negotiating texts as organizational tools to raise the tempo of the process. As discussed in Chapter 11, he issued three versions of the evolving negotiating text as conference room papers (CRPs), each recording and codifying progress made to date, helping tentative agreements to become permanent. Estrada also used his CRPs to circulate provocative text on difficult issues, which then spurred parties on to address those issues in a more concerted fashion. In the second of his three conference room papers (FCCC/CP/1997/CRP.4), for example, Estrada circulated a proposed list of emission targets for each Annex I party. This had the desired effect of prompting many of those parties to speak to Estrada personally about their targets, giving him a much better idea of what they could and could not accept.

The three conference room papers were also instrumental in enabling Estrada to achieve a useful sequencing of the final negotiations, whereby he explicitly attempted to deal with 'seemingly more technical issues . . . early on during COP 3' (FCCC/TP/2000/2, p37), such as the question of whether to adopt multi-year or single year targets, methodological issues and provisions for EITs. These were all agreed in principle – albeit not easily – in the first week of negotiations, and the inclusion of corresponding text in FCCC/CP/1997/CRP.2 – the first of the three conference room papers – made it much less likely that the agreement would be reopened. Estrada also placed strong pressure on the contact groups on sinks and the EU bubble to reach agreement at an early stage, as he knew that many Annex I parties could not decide on their targets without knowing the outcome of negotiations on these two issues. He therefore focused negotiating time on these issues and, thanks also to the deadlines and verbal exhortation he imposed, both texts were indeed preliminarily agreed by the start of the second week.

On several occasions, Chair Estrada overcame the tendency for delegates to delay striking deals by simply gaveling text through. The draft provisions on review of commitments, EITs, and borrowing,[3] for example, were gaveled through (or, in the case of borrowing, deleted) in this way and then codified into FCCC/CP/1997/CRP.2, with Estrada stating that they could be revisited later as part of the package. However, he later strongly discouraged these issues from being reopened. Estrada's actions helped to avoid a situation where all issues – including those of lesser importance – were backloaded to the second week.

This set the scene for a feasible – if very difficult – process of final deal-making on the last night of negotiations. Most of the actual drafting work had been done by that time, and 90 per cent of the text already enjoyed preliminary agreement. Only a small amount of text – albeit very important clauses – was actually changed in the final plenary. It was basically only a handful of clear-cut political decisions on the most controversial issues that were made at the last moment – notably whether or not to include emissions trading, and/or voluntary commitments, and/or the CDM, and the level of target for each party. The process was still one of negotiation by exhaustion – with all its implications for transparency, procedural equity and quality of the text (see below) – but it was manageable.

Time management in the post-Kyoto negotiations

The first two stages

With the benefit of hindsight, it is clear that the time management of the first and second stages of the post-Kyoto negotiations was problematic. Some issues – notably compliance and LULUCF – did follow a relatively coherent process passing through the first two stages of the negotiations – with exploration followed by the tabling of proposals, leading to a tentative transition to bargaining – as reflected in the development of their negotiating texts (see Chapter 11). Other issues, notably the flexibility mechanisms and adverse effects, appeared to remain stuck at the stage of exploration and tabling of proposals, without any transition to bargaining prior to (or even at) COP 6. Negotiating texts either swelled or stagnated. On most issues, there was a sense that the positions of parties became more, rather than less, entrenched over time as COP 6 approached. Unofficial contacts among delegations behind the scenes, although they occurred, simply did not yield the same kind of constructive groundwork and promise of eventual compromise that was so important to the success of COP 3. Grubb and Yamin note that 'what was unusual about The Hague was the extent to which initial national positions hardened and became more extreme as the summit got nearer. This tendency was exhibited by all sides' (Grubb and Yamin, 2001, p267). In addition, the number of issues, options and proposals under consideration mostly continued to swell – rather than streamline – even in the latter months prior to COP 6, at which point the honing down and focusing of issues should have been underway. Ott (2001, p284) recalls: 'The delegations for the most part behaved as if there were plenty of time…when in fact time was already extremely short'. The result was that COP 6 was faced with a vast, complex, interlinked agenda, with little sense among delegates of how the various key issues and options – which themselves had not been spelled out – might fit together, and indeed how they would be addressed in organizational terms. According to Jacoby and Reiner, 'The three years from Kyoto to The Hague were frittered away, leaving negotiators with more issues outstanding at the opening of COP 6 than at the end of COP 3' (Jacoby and Reiner, 2001, p301).

The explosion of issues, and the procrastination and intransigence of parties, was, of course, largely due to wider substantive and political factors rather than the way in which the negotiations were organized. However, it is arguable that time management strategies by the organizers of the negotiation process did not help to counter these tendencies as much as they could have done. One of the major factors here was the disaggregation of the negotiations into multiple negotiating forums, presiding officers and secretariat teams, so that there was no single team of organizers who could take an overview of the whole span of the negotiation process – from COP 4 to COP 6 – and identify steps needed in the negotiation process and their timing. Where this was done on individual issues, for example in the LULUCF negotiations, which followed an agreed decision-making process and defined timetable (set out in FCCC/SBSTA/1999/5), the result was a more structured process with less backloading. Another set of issues on which a clear work programme was followed, to good effect, were the negotiations on guidelines for national systems, reporting and review (Articles 5, 7 and 8) under the Kyoto Protocol. Here, the SBSTA endorsed a work programme prepared by the secretariat (FCCC/SB/1999/2), including the sequencing of issues, whereby guidelines for national systems were tackled first, before guidelines for reporting and review. Although the negotiations were not able to fully adhere to the timetable in the work programme, the sequencing of issues did mean that guidelines for national systems were agreed as early as SBSTA 12 in June 2000, so that at least one issue was off the COP 6 agenda.

Ironically, a factor that may have encouraged foot-dragging by parties was the last minute success of COP 3 and the belief that it could be repeated. Parties seemed to assume that they could continue to procrastinate until the last moment, and still reach agreement. This, however, reflected a lack of understanding of all the important groundwork that had been done in the period before Kyoto, and that needed to be done for COP 6 too.

There was certainly no lack of rhetoric on the part of the subsidiary body Chairs in their public statements to try to push parties to intensify their negotiations. However, probably because they were not responsible for the negotiations as a whole, the presiding officers did not engage in the forceful bilateral pressure on key parties that Chair Estrada had pursued during the Kyoto Protocol negotiations.

Negotiating arenas and negotiating texts were not systematically and strategically used to engineer watersheds that would spur parties on to pick up the pace of their negotiations. The negotiations began in contact groups convened by the subsidiary bodies in 1999, and those same contact groups were used throughout, including in the first week of COP 6. Negotiating texts on all issues did evolve over time, albeit to differing extents, but they were not billed as watersheds in the same way as they were during the Kyoto Protocol negotiations. Moreover, as discussed in Chapter 11, there was no consistency between the textual development paths followed by the various issues until COP 6 itself. The progression from text to text simply did not induce the same sense of urgency in delegates when it was separated out between several issues.

The post-Kyoto negotiations did not have any natural break akin to the six-month rule of the Kyoto Protocol negotiations. COP 5, however, did have the potential to act as a symbolic marker to intensify the pace of negotiations as COP 2 had done. To this end, the organizers convened an 'informal exchange of views' among ministers and encouraged adoption of a decision on the intensification of the process. Many commentators did see COP 5 in this light. ENB stated that COP 5 had generated '"an unexpected mood of optimism" among delegates and observers' with the informal exchange of views 'launch[ing] a year of intensive high-level engagement in the run up to COP-6' (ENB, 1999, p14). Parties also commented on the 'spirit of understanding' that prevailed at COP 5 (ENB, 1999, p14). This feel-good factor, however, was largely due to the lack of political content at COP 5, and the simple deferral of any difficult issues to COP 6. Ironically, therefore, it was precisely the well-mannered atmosphere of COP 5 that meant it did not have the galvanizing effect of COP 2 with its confrontational political drama over the Geneva Ministerial Declaration. Indeed, the possibility of drafting a declaration at COP 5 was all but excluded, following concern not to repeat the contentious experience of COP 2. The COP 5 roundtables were thus open to all heads of delegation, and resulted in a simple summary of discussions, which had no impact at all on the political negotiations (on roundtables, see Chapter 13). Delegates at COP 5 thus indulged in cooperative debate and politely agreed to disagree, and were not forced to face up to their differences. The result was that latent controversies were not tackled, and were simply postponed to COP 6 where they erupted with damaging force.

The equivalent of the Geneva Ministerial Declaration that was adopted at COP 5 was decision 1/CP.5, on the implementation of the Buenos Aires Plan of Action. This decision was intended to mark a turning point in the negotiation process, speaking of the need 'to intensify' the negotiation process, and including invitations to the President and COP Bureau to make recommendations on the organization of work at COP 6. The implementation of this decision, however, was weak. Although it did lead to the convening of several friends meetings before COP 6 (see Table 9.3), no agreement on the organization of COP 6 – or indeed any other appreciable formal or informal advance – resulted from them, and there was no discernible intensification of the negotiation process after COP 5. Again, this failure to follow up on decision 1/CP.5 in any meaningful way can be attributed largely to the disaggregation of the negotiations. Another element was the rule whereby the COP President is only elected on the opening day of the session. The President during the preceding year – in this case, the COP 5 President – thus has much less of an incentive to exert every effort to bring the forthcoming COP to agreement.[4] This lack of motivation, along with the rather weak (albeit genial) leadership of the COP 5 President, also contributed to the fact that COP 5 proved less of an effective symbolic marker than COP 2. Indeed, rather than facilitating the process, it was at COP 5 that the agenda for COP 6 became seriously overloaded, as additional issues, such as the impact of single projects (the Iceland question – see Chapter 3), technology transfer and capacity-building were all backloaded to a deadline of COP 6. A strong presiding officer in charge of the whole negotiation process might at least have tried to sequence negotiations, deferring technology transfer and capacity-building until the session after COP 6, or even perhaps agreeing on them at COP 5, but in the absence of such a leadership figure, the option of deferring difficult topics to a common deadline of COP 6 was too tempting for naturally procrastinating delegates. With hindsight, the burden became unmanageable in the time available.

The COP 6 deadline was undoubtedly talked up by all those involved in the process. The Buenos Aires Plan of Action, however, did not even mention a specific deadline, while the wording of its individual component decisions and subsequent COP 5 decisions varied. Some imagination was therefore needed to interpret COP 6 as a clear-cut and firm deadline. The reinforcement of the deadline by the presiding officers and secretariat, however, meant it did indeed become seen as final, although with perhaps a degree less finality than the COP 3 deadline. This may have been partly because delegates were subconsciously aware of the great difficulty in meeting it, and partly because the forum in which negotiations had been conducted – the subsidiary bodies – would continue to meet beyond COP 6 (unlike the AGBM, for example, which was disbanded before COP 3). The timing of the US election was also thought to hamper the ability of the US to take decisions at COP 6, as the government would be in between (possibly different) administrations (although the chief US negotiator in The Hague denied this).

The finale

Instead of countering tendencies to brinkmanship, it is arguable that time management at the COP 6 finale actually exacerbated these, leading to perhaps the most overburdened last night of negotiations in the climate change regime to date. This reflects the fact that the Presidency, which took an active lead in the organization of the negotiation process, appeared to share the implicit assumptions of the parties that a deal would only, and could only, be struck through negotiation by exhaustion at the last minute. While probably true, this neglected the fact that prior groundwork to enable a last minute deal to be struck is absolutely essential.

A second issue that set the scene for problematic time management at COP 6 was the assumption on the part of the Presidency, and also the higher echelons of the secretariat, that political agreement on the key points of the post-Kyoto deal would need to precede technical drafting of its details. This would require strict sequencing and time keeping. As the Executive Secretary made clear in his opening statement:

> substantive results must be achieved in the first week, and the main political agreements in the middle of the second week, leaving enough time for the consequential technical drafting to be completed before closure (see COP 6 report, part I, paragraph 24).

Unlike the Executive Secretary, however, the ambitions of the COP 6 President seemed to extend only to reaching political agreement on the key issues – and, as noted above, he appeared to share the view that this could only be done at the last minute – with the technical drafting either squeezed in at the end, or deferred to a future session. Problematically, the assumption of sequencing of the political and the technical was not generally shared or understood by most delegations. This meant that, from the outset, the Presidency, most delegates and different individuals within the secretariat were operating according to different assumptions as to the timing, the sequencing and even the final output of the session.

The COP 6 President, and Chairs of the subsidiary bodies and informal groups were not shy in using verbal exhortation to try to push parties to intensify bargaining and start to strike deals early on in COP 6. However, this rhetoric was not backed up by other actions on the part of the organizers. The organizers did convene new negotiating arenas and issue new negotiating texts to try to accelerate the pace of the process, but, with the benefit of hindsight, each of these moves was carried out roughly a week too late.

The traditional marker at the start of the finale to signify the launch of intensive bargaining – namely, the convening of new, or intensified, negotiating arenas – was absent at COP 6. Instead, the same negotiating arenas as before – the subsidiary bodies and their informal groups – continued work in the first week, under much the same Chairs. It was not even a separate session of the subsidiary bodies that launched COP 6, but the *resumed* 13th session. It is unsurprising, therefore, that the rather slow-moving dynamics of the previous negotiating sessions resumed too, and limited progress was made. Delegates were simply not used to intensive bargaining and deal-making on such important decisions in the subsidiary bodies. Furthermore, the knowledge that a new set of negotiating arenas would be launched in the second week led delegates to a psychology of believing that intensive bargaining and deal-making could safely be postponed until that second week.

The organizers certainly introduced a very clear break at the start of the second week, with the convening of entirely new – indeed unprecedented – negotiating arenas, including the informal high-level plenary (IHLP) and four ministerial 'cluster' groups (see Chapter 9). By then, however, a week of negotiations had already passed, and with little to show for it. Even the new negotiating arenas did not launch into the final bargaining process with the necessary urgency; by Tuesday evening of the second week, only two of the four ministerial groups had met at all. The next major break in terms of negotiating arenas, the convening of a friends group, again came very late. Convening a friends group can indeed be an effective means of launching deal-making, and was used successfully in this way at COP 6 (part II) and COP 7 (see Chapter 9). However, at COP 6, the friends group was not launched until the Friday morning, with the scheduled end of negotiations just seven hours away.

The very late convening of the friends group was linked to the very late release of the Chair's text, the Note by the President. President Pronk had planned to announce, on Wednesday evening of the second week, his intention to produce a text the following morning (although rumours that he would do so had been circulating for several days). This would have allowed for 48 hours of negotiations before the scheduled deadline. However, he was misled by positive reports of developments in the negotiations from the ministerial group Co-Chairs and delayed his announcement until Thursday morning. With hindsight, this was a mistake – the groups that met on Wednesday night seemed to backtrack on any progress they had made, or perhaps the ministerial Co-Chairs, when they next reported to President Pronk on Thursday morning, simply abandoned diplomatic pretence. Whatever the explanation, the prognosis on Thursday morning was much dimmer than it had been on Wednesday evening. By then, 12 more hours had been lost.

On Thursday morning, President Pronk then announced that he would prepare a President's text and circulate it at around 4pm. This had the unfortunate effect of stifling an entire day of work, as delegates simply waited for the text. Unsurprisingly, the paper came out late, not until 19:45. A whole day of negotiation, the second to last day before the deadline, was therefore lost. The novel approach of the text, as discussed in Chapter 11, meant that even more time was needed for parties to make sense of, and consider, its substantive contents and their implications. Its lateness was therefore all the more serious.

As in COP 3, the arrival of ministers was used as an important symbolic marker to intensify negotiations, with President Pronk repeatedly appealing to parties to ready themselves for the ministerial segment. President Pronk's strong reliance on ministerial input made this marker all the more important. However, in addition to difficulties surrounding the whole issue of ministerial input (see Chapter 13), timing factors also intervened. One important factor was that ministers arrived earlier than usual, that is, a whole working week before the scheduled end of negotiations, rather than just for the last two or three days. This period may have been simply too long (Yamin and Depledge, 2004). Parties did not feel ready to engage in final deal-making early in the second week when the ministers arrived, so that the first two days of ministerial presence were spent in essentially bureaucratic negotiations, where 'not much was achieved to justify the expenditure of two days of precious negotiating time' (Ott, 2001, p281). By the time they should have started such political deal-making towards the end of the week, ministers were already tired, the natural burst of political energy generated by their arrival had faded, and some ministers even had to leave, as a week is a long time for them to devote to one issue away from home.

Another timing issue is that there was no opportunity for rest for either the President, the secretariat or many delegations on the middle Sunday (or indeed at any other time) due to the scheduling of a ministerial meeting on that day. This meant that the organizers in particular, including the President and his team, were going into a full second week already feeling weary, which most certainly would have impaired judgement.

An important contribution to the slow pace of negotiations and procrastination that characterized COP 6 was the failure to unswervingly uphold deadlines. This applied both to the overall deadline of COP 6, and to several issue- or process-specific deadlines that were imposed on various groups during COP 6 itself. The division of the negotiations into two segments – the first and second weeks – itself created an artificial deadline halfway through the COP 6 process. President Pronk then decided to extend the final deadline he had given to the subsidiary bodies from Friday evening to Saturday lunchtime, allowing the various informal contact groups to meet again on Saturday morning. Inevitably, the contact groups sought even more time to meet on Saturday afternoon, as a lunchtime deadline is never considered final. The timing of the closure of the subsidiary body sessions thus slipped further and further back, and eventually into late Saturday evening. The failure to meet this process-specific deadline early on during COP 6 conveyed the message that working deadlines were not necessarily final, and the organizers could be persuaded to extend them. This was not helpful to countering procrastinating tendencies.

Even more seriously, the organizers failed to uphold the unassailability of the COP 6 deadline. Even in the first week, President Pronk suggested that work might not be completed in The Hague. Although he did not repeat this message publicly, it was enough that the suggestion of a resumed COP 6 had been aired at all. If the presiding officer of the negotiations does not wholly believe in the deadline, then the deadline is no longer. Moreover, on Thursday of the second week, President Pronk announced in the IHLP that the deadline was really Saturday, and not Friday evening, the scheduled end of COP 6. Although parties were perfectly aware that facilities would be available to continue negotiating for some hours beyond the scheduled end of COP 6, for the President to admit so was tantamount to implying that the negotiations could logistically go on a day later than that, which was not the case.

The assumption that the final deal could only be struck under extreme time pressure and conditions of negotiation by exhaustion (see below) is almost certainly what prompted President Pronk to keep pushing the real negotiations further and further back during the final day of COP 6. He convened his friends group only at 11am, where he heard general (mostly critical) comments on his President's text. In a clear example of a high level presence at the COP proving counterproductive, as the valuable time of President Pronk was then taken up in a formal plenary meeting attended by the President of Costa Rica. President Pronk reconvened his friends group again at 5pm, at which point he invited delegates to submit comments on his paper by 9pm, which would be discussed at a further meeting of the friends starting at 10pm. The assumption that the process would go to the wall thus became a self-fulfilling prophecy, and intensive bargaining did not start until the final marathon session of the friends group convened (late) at 11pm on Friday evening. The final negotiations in Kyoto started even later on the last night, but at COP 6, negotiators were trying to conclude work on a greater number of far more complex questions, without the benefit of a clean, accepted text. It was simply impossible to do this at the last minute. Parties were excessively complacent in believing that the final deal would be struck at the last minute, without carrying out the prior work needed for that to be possible. Importantly, they were not pushed into doing so by the organizers of the negotiation process. When the crunch came, there was simply too much to do and too little time in which to do it.

A deal – or rather, almost a deal – was forged on the Saturday morning between certain EU countries and the US meeting unofficially behind the scenes (see Chapter 9). However, timing factors again intervened. Even if sufficient time had been available to consult with countries and coalitions not part of the unofficial negotiations, and to transcribe the deal into legal text, there was simply not enough hours left to produce the text as a document, translate it into six languages, and make it available to all delegations for formal adoption before the conference centre had to be dismantled and developing country delegations on UN-funded tickets had to leave for the airport. Negotiations at COP 6 simply ran out of time, but due to poor time management where the tendencies of parties to procrastinate were indulged and exacerbated, rather than to any inevitability.

Negotiation by exhaustion

'Negotiation by exhaustion' is often used to refer specifically to the final 24 to 48 hours of negotiations, where delegates and the organizers work round the clock, often without rest or sustenance, to secure a final deal (or indeed admit failure), almost invariably after the scheduled end of negotiations. This final marathon session typically comes at the end of an increasingly intense week or more of negotiations, where formal, informal and unofficial meetings have also gone on late into the night (Yamin and Depledge, 2004).

The finales of both the Kyoto and post-Kyoto negotiations took place through a process of 'negotiation by exhaustion', escalating into a final overnight marathon session that eventually concluded nearly a day late. In Kyoto, the final 'marathon session' in the Committee of the Whole lasted from 1am to 10:17 on 11 December. The formal adoption of the Kyoto Protocol in the COP plenary did not take place until around 3pm later that day, nearly 24 hours after the scheduled end of the conference. The sequence of events at COP 6 was similar. The friends of the President met for what was intended to be a final round of deal-making at 11pm on 24 November, until around 9am next morning. Unofficial negotiations between the EU and US continued in the morning, but in vain. The final COP plenary ended at around 6pm. Negotiation by exhaustion, including overnight negotiations and late running, also characterized the finales of the political segment of COP 6 (part II), and COP 7.

There is a widespread belief among negotiators, not only in the climate change regime but also in most negotiation processes, that the final deal can only be struck under such conditions of exhaustion.[5] Delegates are inevitably less resistant to pressure, their resolve weakens, and they are more likely to back down from their positions as physical tiredness take its toll. One interviewee commented:

> if we hadn't been under that kind of pressure, under those kind of circumstances, we probably wouldn't have come out with an agreement... You basically have to lift yourself up... work to that kind of pace, to that kind of level, to basically weaken people down so they will consent to compromise.

At a more personal level, the exhaustion of late night negotiations can provide a sense of drama and occasion that some delegates relish. As one interviewee put it:

> at the end of the day... I don't think you can avoid the fact that the show has to happen on the final night... we would like to believe that we are all much more rational, but in fact... it's a psychological thing... We thrive on it.

Delegates must be able to demonstrate that they made every effort to maintain their position before being forced to give in, while late night negotiations can generate a sense of common and shared hardship (Yamin and Depledge, 2004).

According to one interviewee, 'it's the way that diplomats function...Delegates need to feel they have suffered...to be able to say we tried really hard, we sweated, we did our best.' Another interviewee agreed: 'I think that it's a pathology of the individuals involved, in some sense they wouldn't feel that they had done their job properly if they didn't push themselves to late at night'.

While perceived as inevitable, negotiation by exhaustion has very high costs attached to it. Firstly, it does not always succeed. Although exhaustion and the imminent threat of failure can lower the resistance of parties and induce them to compromise to some extent, these factors will not usually cause parties to go beyond their bottom lines. Therefore, if insufficient creative work has been done beforehand to explore possible areas of compromise, then tiredness and time pressure alone will not forge agreement. In addition, tiredness will affect the ability of individual delegates to think through possible compromises and their implications. Nothing new should therefore be introduced on the last night, and parties should simply have to choose between a limited number of options and combinations of options. Again, for this reason, the document before parties must be straightforward and easy to read, so as not to waste scarce mental energy.

The difference between COP 3 and COP 6 in this regard is clear. The final meeting of the Committee of the Whole at COP 3 had a new document before it, but almost all the text was well-known, and most of it was agreed in principle. The choices facing delegates were relatively straightforward and demanded little analysis, except for the implications of the combination of choices. The final friends meeting at COP 6, however, had before it the President's Note which, while released 24 hours earlier, was still extremely innovative in style, content and structure. Moreover, participants in the friends group also had before them a mass of written comments on the text (see Chapter 11), which they were being asked to sift through and somehow reconcile into amended text. The intellectual capacity needed to do this meant that it was simply not appropriate for a process of negotiation by exhaustion.

Negotiation by exhaustion can therefore backfire. An excellent illustration of this point concerns the former French Environment Minister, Dominique Voynet, who presided over the EU at COP 6. Following the conclusion of the apparent last minute deal between the US and certain EU countries, Minister Voynet declared herself too tired to fully understand the deal (especially as it was written only in English) and therefore unable to explain and sell it to the wider EU. Although it is unlikely that the deal would have stuck anyway, this is a clear case of where last minute negotiation by exhaustion placed obstacles to advancing the process.

Not all interviewees agreed that negotiation by exhaustion was a significant factor on the last nights of negotiations. One, for example, from a well-resourced OECD country, said 'tiredness was pretty marginal. We were on top, in command, the adrenaline was flowing.' Interestingly, a non-Annex I party delegate took a similar viewpoint, stating 'I don't think it affects people's ability, because they still have the three key things that they are looking for...despite their weariness'.

Nevertheless, negotiation by exhaustion at the finales of both the Kyoto and post-Kyoto negotiations, including COP 6 (part II) and COP 7, meant that the commitments of states under international law were being negotiated late at night by individuals who were often suffering from extreme tiredness. These were not propitious conditions for taking such important decisions. As one interviewee noted, 'the late night meetings are a terrible way to make public policy, at three in the morning by people who have not slept for three days'.

The tiredness of negotiators can certainly affect the quality of the agreement, as parties are less willing to devote attention to language and style, focusing simply on closing the substantive deal (Yamin and Depledge, 2004). According to Werksman, 'marathon sessions can undermine the quality of decisions, as negotiators . . . fail to choose their words carefully or to ensure the consistency of the text' (Werksman, 1999, p12). One interviewee recalled:

> we nearly adopted a sentence without a verb in it! These things should not happen. It was time pressure. Working 30 hours, and then another 30 hours . . . with hardly any sleep in between, is not a good way to keep your thinking powers intact.

The various discrepancies that appeared in the Bonn Agreements adopted at COP 6 (part II), for example (see Chapter 9), can largely be attributed to the tiredness of delegations and the organizers.

In the fog of fuzzy thinking that characterizes late night negotiations, differing interpretations can even arise as to the text adopted. The decision on adoption of the Kyoto Protocol, for example, led to an inexplicable incident in which, during the technical review of the Kyoto Protocol a month after the close of COP 3, a senior delegate from an Annex I party claimed he was under the impression a different version had been adopted, including text on compliance and more detailed work on emissions trading. Although the large delegation of this party was able to cope better than most with negotiation by exhaustion, it had not been immune from the general 'blur' of late night, intense talks.

A major failing of negotiation by exhaustion is that it hits the smallest and least-resourced delegations hardest (Yamin and Depledge, 2004). Larger delegations are able to establish a rota system, so that the negotiations are constantly covered by a relatively well-rested individual. Such rotation is not possible, however, in small delegations, with serious implications for practical procedural equity. An African interviewee recalled the final night of negotiations at COP 3 thus:

> at a certain point, I fell asleep. And what is said while you're asleep, it's not guaranteed that you'll be happy with it . . . once again, it's the small delegations that suffer. If there are ten of you in a delegation, five can sleep, and five can take over. The others can go and have a rest . . . But it's a real problem for us.

Another African interviewee echoed this sentiment:

Coming from the South, I definitely and vehemently oppose late
night meetings since we do not have the numbers to sustain it and
therefore they do not work in our favour...those late night
meetings appear suspicious in intentions and lack good faith.

In the case of COP 3, the overrunning of the negotiations also meant that
interpretation facilities were lost before work had concluded, placing non-
Anglophones, especially less well-resourced developing countries and EITs,
once again at a disadvantage. In addition, many negotiators, especially
developing country delegates, were forced to leave the conference centre
before the close of the negotiations to catch their flights home. In the case of
the three finales of the post-Kyoto negotiations – COP 6, COP 6 (part II) and
COP 7 – the final marathon round of deal-making in any case took place in
informal, English-only arenas, effectively excluding non-Anglophones. The
problem of premature departures from the conference centre was, however,
minimized to a large extent by the secretariat having anticipated the problem
and booked later flights for funded delegates.

Summary and concluding remarks

The messages of this chapter are straightforward. Judicious time management
over the whole course of a negotiation is very important to establishing the
conditions that can enable agreement to be reached within the deadline. While
there is a natural tendency among negotiators to backload negotiations, it is the
task of the organizers to counter this tendency. This is easier when there is a
single team in charge of the whole process from start to finish and across
different issues, enabling forward planning and strategic time management.
Convening more intensive negotiating arenas and more advanced negotiating
texts, as well as talking up natural break points (e.g. mid-point COPs) and
unswervingly upholding the deadline, are all important ways of pushing parties
to move forwards and not get stuck at a particular stage of the negotiations.
Managing the final negotiating session and constantly challenging the tendency
to brinkmanship is particularly important.

It is almost inevitable that the final deal will be struck at a late night, last
minute marathon session, which will almost undoubtedly overrun past the
scheduled closure of the Conference. For such negotiation by exhaustion to be
successful, however, sufficient ground must already have been covered. Even if
successful, negotiation by exhaustion has damaging repercussions, on
procedural equity, for example, on transparency, and on the quality of the
agreement. The inevitability with which many negotiators view this factor can
be largely attributed to the prevailing perception of negotiations as
confrontational bargaining, rather than exercises in joint problem-solving.

The Political and the Technical: Ministerial Input

The start of the ministerial segment...the moment at which the Great and the Jet-lagged join the Wise and the Weary (GLOBE, 1998).

Introduction

This chapter explores arrangements for the participation of ministers in the climate change negotiations. A distinction is typically made in both the literature (e.g. see Wettestad, 1999) and policy arena between administrative-level civil servants – termed officials – and higher-level political participants representing the government in office, known as ministers[1] in most countries. These different actors are viewed as having contrasting strengths. While 'bureaucrats [officials] often master the complex technological and political details...ministers are...freer to cut bargains, and they usually draw much more media and public attention to the issues' (Wettestad, 1999, p23). A particularly important contribution that ministers can make to the negotiations is one of providing political leadership, or skill and energy; that is, being able to articulate and implement a broader vision than officials who have no authority to stray from the government line, and thus help to forge an agreement based on the work of those officials. An effectively organized negotiation process should thus seek to draw out a synergistic, productive interplay between ministers and officials and their differing strengths. In this chapter, we explore the four main avenues in which ministers have been involved in the climate change negotiations: the traditional 'general debate'; roundtable forums; direct involvement in the negotiations; and unofficial dealings behind the scenes.

The high-level segment

Each COP session to date has featured a so-called 'high-level segment' or 'ministerial segment' intended for participation by ministers. The high-level segment has almost always been held on the last three days of the COP session,[2] with the intent that ministers would provide important political input into the most difficult decisions needed in the final deal-making stage of the negotiations, building on the work of the more technical officials. Ministerial presence at COP sessions has always been high, especially among Annex I parties, who are almost all regularly represented by ministers. The proportion of non-Annex I parties represented by ministers is typically lower, although it has been rising. This immediately points to an important underlying issue, that is, differences in the extent and mode of participation by ministers from Annex I and non-Annex I parties. Indeed, concerns on the part of many non-Annex I delegations that they should not face exclusion from any part of the negotiations on account of their not being represented by a minister led to the renaming of the 'ministerial segment' (at COPs 1 and 2) to the 'high level segment' (from COP 3 onwards), therefore also encompassing senior officials (Yamin and Depledge, 2004).

The general debate

The traditional formal general debate is part of established practice for ministerial participation in UN forums. As part of this debate, ministers and other heads of delegation deliver speeches outlining their national positions from the podium in the main plenary room. This debate is chaired by the President, a task that s/he almost always refers to the Vice-Presidents, to enable him/her to get on with work in the real negotiations. Occasionally, heads of state choose to speak at high-level segments, most commonly the head of state of the hosting country, but also sometimes others with a particular interest in the negotiations. The President of Costa Rica, for example, attended COP 3 and COP 7, while the President of France attended COP 6.

High-level segment speeches are almost all pre-prepared statements of a 'highly predictable and rhetorical nature' (Werksman, 1999, p13), with the order and timing of delivery tightly choreographed by the secretariat. Participants thus very rarely respond to one another and the debate in fact typically consists of monologues. Despite the potential for ministers to use the formal debate as a platform for exerting inspirational leadership, in practice this rarely takes place in any meaningful way. Except for ministerial statements from the most influential countries, which are listened to avidly for signs of any change in position, there is no expectation that the formal debate will feed into the negotiations. The speeches are generally not summarized in the report or recorded in any way,[3] except for a listing of speakers in the COP report. The exception was at COP 5, where statements were placed on the UNFCCC website, and the ENB was commissioned to prepare an index of key topics covered by the statements. This was an explicit attempt by senior staff within the secretariat to increase the impact of the general debate. A hard copy of the text of most speeches can be obtained from the secretariat.

A general debate has taken place at each COP session up to and including COP 7.[4] The number of speeches delivered has varied between 75 (at COP 7) and 125 (at COP 3). The high-level segment is one of the very few cases where the COP acts to limit the speaking time of delegates. From COP 1 to COP 4, speakers were granted five minutes, but since then the time limit has been set at 3–4 minutes. The order of speeches – the list of speakers – is maintained by the secretariat, with slots given on a first-come-first-served basis, but with priority to ministers. Prime slots are highly prized.

Various methods have been tried to increase the efficiency of the general debate and minimize the instances where meetings have gone on into the night and speakers have addressed an almost empty hall. Stop clocks and even bells have been introduced to encourage speakers to stick to their allotted times, although the determination of the chairing President/Vice-President to enforce time slots has varied, as ministers rarely take kindly to having their words curtailed. Parties have also been strongly encouraged to deliver statements as a group (e.g. AOSIS) rather than individually, with the incentive of a better and longer speaking slot. Another tactic has been to suggest that countries refrain from speaking, and instead circulate paper copies of their statements.

The general debate does fulfil some valued functions. One such function is inclusivity, ensuring that all heads of delegation from all countries have the opportunity to participate and be seen to participate, albeit passively, in the negotiation process. In doing so, the formal debate can help raise the political, public and media profile of the negotiations, while conferring greater status on the climate change issue on the international agenda. The high ministerial presence at COP 3 for example, and, in particular, the presence of the US Vice-President and the heads of state of Costa Rica, Japan and Nauru directed the spotlight of the world's media firmly onto Kyoto. Raising the profile of climate change can be similarly important at the national level, especially for developing countries, where the issue tends not to be so high on the political agenda (Yamin and Depledge, 2004). The attendance of a minister or head of state at a COP, and the delivery of a speech to the plenary, can become a news item in that country and thereby increase public awareness of climate change.

More substantively, the formal debate can become an important vehicle for key players to announce major new positions or pledges, which can then impact profoundly on the negotiations. At COP 2 in 1996, for example, the US Under Secretary of State for Global Affairs made a dramatic statement to the effect that the US would support legally-binding targets, within the context of an emissions trading system, while attacking climate sceptics. This marked an important turning point in the Kyoto Protocol negotiations. Even more dramatically, at COP 3, the US Vice-President, Al Gore, departing from the printed text of his speech, announced that he was instructing his negotiators to 'show increased negotiating flexibility' (cited in Oberthür and Ott, 1999, p86). Similarly, announcements by countries – notably again the US – during the high-level segment at COP 4 that they intended to sign the Kyoto Protocol were very important to shaping the mood and outcome of that session. Argentina and Kazakhstan's announcements, at the same high-level segment, regarding their intentions to take on voluntary targets also had important repercussions on subsequent negotiations. Although the substance of these statements cannot be attributed to the formal debate as such, the convening of a suitable forum attracted high-level participation and provided a high profile backdrop that added dramatic effect to the statements made, ensuring that they were well-publicized and made in the most formal way possible.

In most cases, however, the traditional general debate, despite the high political profile of its participants, has taken place on the margins, rather than at the centre, of the negotiation process, with a real disconnect between the two. Many interviewees, while recognizing the contributions made by the general debate, doubted whether the costs in terms of negotiation, interpretation and secretariat time were worth it. According to the senior secretariat official speaking in his interview, 'it is a heavy price to pay'. The secretariat did put forward proposals to the SBI for alternative means of involving ministers as early as 1997 (see FCCC/SBI/1997/11). However, the secretariat's suggestion to do away with the formal debate was resisted by many parties, especially developing countries, but also the US, which was likely anticipating a visit to COP 3 by its Vice-President.

A good illustration of the rather ambivalent attitude of participants in the climate change regime to the general debate can be found at COP 6. Here, facing strong time pressure, President Pronk appealed to ministers to forego some of the interpretation time allocated to their general debate, to instead allow the IHLP – the main working body of the negotiations – to meet. Although many parties and their ministers concurred, others were very unhappy at what they saw as a political snub, with the objections of the Russian Federation turning into a minor diplomatic incident. It was not until preparations for COP 8 that parties agreed not to hold a general debate for that session, and instead to channel input by ministers through more interactive 'roundtables'. We now turn to examine these forums.

Roundtable forums

The first ministerial roundtable in the climate change regime was held for half a day at COP 2. It enjoyed only mixed success, partly due to attendance being limited to **heads of delegation of ministerial rank** (see FCCC/CP/1996/1/Add.1, emphasis in original). The rationale was that ministers would be more likely to engage in substantive discussion with their peers than if officials were also present. This approach, however, while justified in terms of efficiency, caused widespread discontent on the grounds of procedural equity and transparency among delegations not represented at ministerial level. One AOSIS interviewee recalled:

> AOSIS was only given limited access…because we hadn't brought any ministers, we had considered it to be a non-ministerial meeting, [my country] was only represented by myself and the Assistant Attorney General, so we weren't even allowed in.

The ministerial declaration that emerged from the roundtable (and associated behind the scenes negotiations) was rejected by several parties and could not be adopted, partly due to objections over lack of transparency in its development.[5]

A second experiment at COP 5 – dubbed an 'informal exchange of views' – open to all ministers and heads of delegation proved much more acceptable, and paved the way for acceptance by parties that a roundtable forum would be the main means of eliciting ministerial input at COP 8. A similar roundtable was held at COP 9 and is planned for COP 10. The roundtables held to date in the climate change regime, and their main features, are summarized in Table 13.1. It is worth highlighting from the outset that roundtables have been held at quieter COP sessions, that is, not those serving as finales of negotiating rounds where a set of important decisions must be taken. It was expected at COP 3, COP 6, COP 6 (part II) and COP 7 that ministers' time would be entirely occupied with the negotiations themselves. This immediately points to the role of roundtables as time fillers, an issue we will return to later in this chapter.

The main aim of the roundtables is to promote debate among ministers that is more interactive and frank than the monologues of the general debate, in the hope that ministers will exchange views, develop their own understanding, and learn from each other in ways that will eventually feed into a more constructive negotiation process (Yamin and Depledge, 2004). To this end, the roundtables are designated as *informal* forums, underlining that words spoken do not imply any commitment, and can therefore be shared more openly. Since the ambiguous experience of the COP 2 roundtable and its ministerial declaration, no attempt has been made to actually negotiate anything in, or on the margins of, these roundtables. Instead, summaries of discussions and key points raised have been published in the relevant COP report, with each session showing greater boldness in the extent of detail included. The COP 9 report, for example, included a full page or more on each of the three roundtables held.

The conduct of proceedings in the roundtables does, in fact, have much in common with the traditional general debate. All ministers and heads of delegation have the opportunity to participate and make a statement, and the events are held not around a round table, but in a plenary room. Considerable efforts, however, have been made to try to distinguish the roundtables from the general debate, and promote lively and focused debate. For example, each roundtable has been based on a defined theme. These themes have inevitably been quite broad, given that they have been negotiated and agreed among parties in the SBI. However, they do represent a step forwards from the random, unrelated statements that make up the general debate. Another strategy has been to invite a small number of parties to serve as 'lead off' speakers to put forward ideas and get the discussion rolling. Appointing ministerial Co-Chairs for each roundtable has also been important in bringing buy-in from other ministers, not just the COP President, encouraging them to give the roundtable their own personal touch. The absence of a list of speakers means discussions are less closely staged, and seating has been arranged in different formations to convey a more relaxed and intimate atmosphere. The main NGO constituencies have been represented and allowed to observe proceedings, and in some cases also to make statements.

The extent to which the roundtables have succeeded in provoking open, substantive debate among participants is limited (Yamin and Depledge, 2004). Although some ministers have made spontaneous interventions and reacted to other speakers, the majority has still delivered prepared statements and actual discussion with off-the-cuff remarks has been rare. As a secretariat official explained, 'everyone is pro "free flowing discussions" etcetera, but still there are prepared speeches, officials still want to know exactly when their minister will be speaking, on what, for how long'. One means of seeking a middle way between the stiltedness of scripted statements on the one hand, and the wariness of ministers (and their staff) of entirely spontaneous debate they might be unprepared for on the other, has been for the President to circulate an advance 'survey' to delegations. This survey, circulated at COP 8 and COP 9, invited delegations to indicate their advance preference of which roundtable they preferred to contribute to, along with 'key words' of their planned contribution. The debate at the roundtables was then organized in advance by the presidency and secretariat based on responses received, so that proceedings were loosely structured, but not as tightly choreographed as the general debate. Several interviewees also pointed to the skilful chairing by many of the ministerial roundtable Co-Chairs at COP 9 who actively sought to promote real discussion, albeit with limited success. One interviewee calculated there had been 17 off-the-cuff statements in the technology roundtable at COP 9; the fact that he was able to actually identify the number of such statements suggests they were not as off-the-cuff as all that. According to the interviewee, most of these were made by OPEC parties, who do not need encouragement to make uninhibited interventions. Another interviewee commented: 'In the end it hasn't changed a lot, they [the ministers] still make a speech for their domestic audience. Trying to focus them on topics has not been successful.'

Although most statements remain scripted, the roundtable forums have nonetheless provided a helpful environment for ministers to speak 'outside the box', that is, to air views on wider subjects than the strict negotiating agenda, while also making linkages between issues. This indeed reflects one of the strengths that are expected of ministers – to go beyond detail into the broader issues. In a similar way to the special events convened mostly by NGOs (see Chapters 10 and 14), the ministerial roundtables provide an opportunity for ministers to raise topics that are not being discussed in the negotiations themselves, to showcase national initiatives relating to climate change implementation, or to make national pledges. In its analysis of COP 9, the ENB reported:

> outside the box of defensive party positions...the high-level round-table discussions among ministers provided a refreshing change of pace, allowing an opportunity to step back and take a wider perspective on the UNFCCC process...unleashed from common denominator group positions and the confines of negotiations.

Although an improvement on the general debate, the merely modest achievements of the roundtables in actually delivering meaningful interaction can be attributed to a number of factors. At a basic level, the size and open-ended nature of the forum is problematic. As noted in Chapter 9, in-depth debate in a large plenary format is rarely possible. One interviewee put it thus: 'It's got to be done in a closed room, over dinner. It can't be done in plenary.' More fundamentally, the actual aims of roundtable forums are problematic, and to some extent contradictory. Ministers are asked to engage in free-flowing discussions and to feel they can speak openly because the forum is informal and no bargaining will take place, yet if no bargaining takes place – not even on a joint declaration, statement or report – then there appears to be little point or clear direction to the discussion.[6] Mutual learning and the sharing of experiences are important, especially across developed and developing countries, but as one interviewee put it, 'ministers come [to the climate change negotiations] to make decisions, not to share information'. A fundamental problem is that, as with the general debate, there is little connection – aside from within national delegations – between the ministerial roundtables and the negotiations proper. In many ways, therefore, the ministerial roundtables are as much of a side show as the NGO side events.

In this sense, roundtables have not succeeded in harnessing the political energy of ministers to actually push issues forward. There has been no identification of pressing issues, no statement of common purpose, not even a message conveyed to the negotiating officials. Just like the general debate, roundtables are widely viewed as a filler; as one interviewee put it, 'they've kept the ministers off the street... I'm not sure it's a good use of ministers' time for two days'. The impression is of ministers being corralled into a politically benign, time-consuming forum, so that the negotiators can get on with the 'real' work. Another interviewee stated that he would find it difficult to persuade his minister to come to the next climate change COP, if the only opportunity for him to participate in the negotiations was through a roundtable. A recurring message from interviewees was that ministers 'like to feel useful', and the roundtables simply do not make sufficient use of the potential of ministers to unblock, and give energy to, political processes.

Direct participation

The most important impact of ministers on the climate change regime is, of course, through their direct intervention in the negotiations, usually in the very final deal-making stage, when bargaining and deal-making among officials has gone as far as it can, and political, high-level input is needed on the most difficult issues. The role of ministerial input in this regard is widely revered in the climate change process. In his statement to the opening of the COP 5 high-level segment, the Executive Secretary told ministers, without apparent irony, 'Your arrival lifts this conference from tactics to vision' (Zammit Cutajar, 1999). Several interviewees similarly emphasized the importance of having ministers on site to bring the negotiations to closure. As one remarked, ministers are 'the icing on the cake... they are there to solve the problems that officials cannot'. Another commented, 'ministers... don't know the details, and therefore they will not necessarily make the "right" choice ... but they'll make a decision ... they'll reach an agreement'.

While officials have to negotiate according to instructions from their governments and not deviate beyond a certain bottom line, ministers, who are members of the government that has issued the instructions in the first place, can agree to change that bottom line. Moreover, while officials who have been involved in the negotiations for some time tend to have a good grasp of details, including the history of the process and the importance of particular words and phrases, ministers are able to see the bigger picture and relate the negotiations to other policy areas. Indeed, ministers can often be more radical than their officials, as they do not carry the full historical baggage of the negotiations with them. This can cause friction. A Latin American official, for example, recalled how she had stepped in to stop her minister from expressing support for joint implementation with developing countries – contrary to the national and G-77 position – during the Kyoto negotiations. In a different example, the South African Minister, in his capacity as facilitator for the final negotiations on compliance at COP 6 (part II), put forward a radical proposal whereby a party that failed to comply with its emission target under the Kyoto Protocol should make 'reparations' (Lefeber, 2001, p31) for the resulting environmental damage. This was an entirely new proposal coming at a very late stage, and lack of agreement on it was one of the final sticking points in the negotiations on that issue. As a participant in the negotiations recalled, 'the introduction of this new element in the negotiations by ... Valli Moosa [the South African Minister], had a profound impact on the contents of the political agreement on compliance' (Lefeber, 2001, p39). In a similar vein, in his compromise text (the 'Note by the President') issued on the penultimate night of COP 6, COP President Minister Pronk introduced 'two extremely interesting but completely novel ideas with complex international ramifications that had not been discussed in public on even one single occasion during the previous three years' (Grubb and Yamin, 2001, p269). The radical nature of Pronk's text (see also Chapter 11) was certainly a contributor to the failure of COP 6.

Despite the dangers of excessively original ministerial input, the broader approach that ministers can offer is important in the final stages of negotiations when a package deal has to be constructed. Given the economic implications of climate change, particularly for national competitiveness, it is important for ministers from different governments to be able to talk directly and reach understandings with one another. Ministerial presence was therefore absolutely crucial in sealing a deal in the intense final stages of the Kyoto negotiations at COP 3, and the post-Kyoto negotiations at COP 6 (part II) and COP 7. According to one interviewee, 'without a ministerial session, Kyoto would have failed ... because the middle-level bureaucrats ... don't have enough power to close the deal. It takes the ministers to close the deal.' Indeed, both at COP 3 and COP 6 (part II), many delegations had to get final decisions approved not only by the minister present, but also by other ministers, or even the head of state. The Presidents of Russia and the US, the Prime Ministers of Japan and the UK, and the Chancellor of Germany were among those who

Table 13.1 *Ministerial roundtables held in the climate change regime*

COP	Name and (abbreviated) topic	Ministerial (Co)Chair	Special features
2	Ministerial roundtable • New scientific findings and opportunities for action	Switzerland	Ministers only. Presentation by IPCC Chair. Two-page Chair's summary in COP 2 report. No NGOs.
5	Informal exchange of views • Progress made: lessons and challenges • The way forward	President Norway/Uganda	About 1 page summary by President in COP 5 report.
8	Roundtable discussion • Taking stock • Climate change and sustainable development • Wrap up	President/UK President/South Africa President	Total of 94 interventions by parties, 3 by NGOs.
9	Roundtable discussion • Adaptation, mitigation, sustainable development • Technology • Assessment of progress Cross-cutting themes: capacity-building; synergy/ possible future steps; awareness of vulnerability and adaptation.	Japan/Marshall Islands USA/South Africa Mexico/Germany	Two paragraph summary in COP 8 report, part I. One page summary on each roundtable by President in COP 9 report, part I. Certain parties invited to serve as 'lead off' speakers. Statements by 90 parties, 2 observer states, and 3 NGOs.

provided input to the negotiations at COP 3 and/or COP 6 (part II) through mobile telephone calls.

Expectations and the dangers of over-reliance

An interesting dimension to the impact of ministerial presence is the influence of expectations, with a deeply engrained perception among delegates that it is ministers who will resolve the outstanding critical issues and bring the process to closure. As one interviewee explained:

> There is still this working myth that we [officials] are the...drones that work to prepare the clear options that ministers will sit down and sharpen their pens and tick the ones that they want...whenever an issue floats beyond a delegation's ability to resolve it, they say 'well, this has now become an issue for ministers'...Ministers rise above what the drones do...Everyone else is operating under instructions, but ministers, they write their own instructions...Whether that myth is true or not, it's one of the negotiating techniques.

This expectation, however, does hold the danger of inducing over-reliance on the part of officials that ministers 'who yesterday were thinking of something totally different and tomorrow will think about something totally different again [would] suddenly come in and...crack two or three seemingly uncrackable points' (interview). As the Russian delegate stated on the eve of COP 3, 'there is a view that the ministers [will] come in next week and decide everything. We should not leave everything to them...this would be a mistake' (AGBM 8, 1997f). This heavy reliance on ministers can be seen as part of the tendency to brinkmanship and the backloading of negotiations, as discussed in Chapter 12.

Effective ministerial decision-making is in fact critically 'dependent on bureaucratic legwork' (Wettestad, 1999, p213). That is, it is important for the negotiating officials to resolve the bulk of the issues, leaving only a small number of political questions for ministers to consider. A stark illustration of this point is the contrast between COP 3 and COP 6. By the time ministers made their debut on the COP 3 stage for the last three days of negotiations, much of the draft language of the emerging protocol had already been agreed in principle, leaving only a few clear decisions for ministers to make (see also Chapter 12). A good example is the draft article on policies and measures. Following intensive negotiations over a complicated text on policies and measures in the first week of COP 3 (see FCCC/CP/1997/2), by the time ministers arrived, the whole text was clean except for one set of square brackets around a single sentence, reflecting outstanding disagreement over the crux of the matter, that is, whether the application of policies and measures should be mandatory or voluntary (see FCCC/CP/1997/CRP.2). The more peripheral, detailed text in the article had been cleared up by officials, allowing ministers to focus their attention on the core political issue that only they could resolve.

The situation at COP 6 was very different. As Grubb and Yamin comment, 'during the first week, there was slow progress on the extensive texts as negotiators ... hung back from making concessions until their political masters arrived' (Grubb and Yamin, 2001, p268). The result was that, when ministers arrived at the start of the second week, they were faced with very lengthy and intricate texts full of square brackets on almost every issue under negotiation. The sheer number and complexity of interrelated decisions still requiring resolution, along with the technical nature of some of these, hindered the process of negotiation and decision-making among ministers, which contributed to the failure of the session.

Developed and developing countries: differing ministerial input

One of the major challenges for involving ministers directly in the negotiations is the much greater activity of industrialized country ministers relative to those from developing countries (Yamin and Depledge, 2004). Although developing countries are usually represented at ministerial level at major COP sessions, their delegations often continue, in practice, to be led by officials, with ministers confining themselves to ceremonial activities. There are, of course, notable exceptions to the rather low profile of non-Annex I ministers. The South African minister, for example, was highly active at COP 6 (part II) and, especially, COP 7, while the Tanzanian minister played an important role in discussions on the LDC fund at COP 9. The profile of the Nigerian minister at COP 6 (part I), where he held the post of G-77 Chair, was also high.

Despite these notable exceptions, general disparity in level of participation between developed and developing countries has raised obstacles to convening effective negotiation forums based on ministerial participation, due to the differing strengths and approaches of ministers relative to officials, as well as issues of protocol. Officials conversant in details can out-manoeuvre ministers not so experienced in technical and textual minutiae, while the lack of power of officials to agree to compromises can frustrate ministers. Several interviewees pointed to this problem. One noted, 'on the Annex I side you get ministerial level, but on the developing country side you get the old cohorts still there ... we found that extremely unhelpful'.

Several developing country interviewees spoke of the problems they faced in securing effective participation by their ministers in the climate change negotiations, including lesser priority given to climate change, language barriers, and even the use of negotiation assignments as patronage. One interviewee claimed that:

> high-level meetings are attended by people who don't necessarily know the issue, who do not necessarily listen. In some countries... the ministers only want to travel...to take advantage of the free ride, so you are not getting very much out of that.

Another summarized the problem thus:

ministers from developing countries have two problems. First, they don't know the issue very well, at least not as good as developed countries. Second, they can't speak proper English ... [My] minister can't speak English at all, so how can he negotiate?

Given these difficulties, it is not surprising that officials continue to be the most active negotiators on developing country delegations. Several interviewees, however, argued that this is changing, albeit slowly, as the profile of climate change grows and developing country governments become more aware of their interests in the issue. The establishment of new funds and the CDM have increased the incentives for developing country governments to participate in the climate change regime, while growing concern over the adverse impacts of climatic changes has also increased awareness of the issue.

Channels for direct participation

Although ministerial input is important for every COP session where key decisions must be made, the ways in which this input is channelled has varied from session to session, with contrasting results. At COP 3, ministers were not expected to participate in the official negotiations ongoing in the Committee of the Whole or informal groups. Although a small number of them were invited to participate in friends groups, these groups never became central negotiating forums (see Chapter 9). Where ministers were particularly active, however, was in *unofficial* negotiations, seeking to broker deals with their counterparts behind the scenes, or sanctioning changes in national positions and approving proposed compromises that went beyond the authority of officials. At COP 3, therefore, ministers were most active in both the most and least formal arenas; that is, almost all ministers made statements in the traditional formal debate, and many were also engaged in unofficial negotiations in the corridors. They were not, however, expected to be involved in the day-to-day negotiating arenas. Ministers could thus play to their strengths, appearing on the plenary stage and delivering monologues on the one hand, and engaging in political deal-making behind the scenes on the other.

The organization of ministerial input at COP 4 differed considerably, largely because the COP President, herself a minister, wished to take a more hands-on role in the negotiations, and therefore work with her peers. At this session, the final deal-making took place in a friends group, where the problems of disparity in participation between ministers and officials were thrown into focus. The Annex I ministers often found themselves overwhelmed by the mastery of technical and textual details on the part of non-Annex I officials, and resented not being able to discuss the bigger picture directly with their counterparts. This contributed to the rather awkward and bad-tempered political dynamics – including a walk out by G-77 delegates from the friends group – which characterized the final negotiations at COP 4.

COP 6 saw these problems magnified many times over. Again, the presence of an active COP President, along with the highly political nature of decisions to be taken, led to an unusually strong ministerial focus in the organization of the negotiations. Ministers were expected to participate in the IHLP (the main COP working body), ministerial Co-Chairs were appointed to head the four informal negotiating groups, and the vital role of ministers was continuously talked up, notably by inviting them to come to The Hague at the start of the second week, rather than just for the last three days. It was certainly true that a strong ministerial presence was needed at COP 6. However, the unbalanced weight given to ministers had unfortunate consequences. Firstly, it alienated the officials who actually knew the issues and the negotiating texts, and, in doing so, upset critical personalities (especially from developing countries) who could have helped forge a deal. Secondly, due to the general disparity in participation discussed above, ministers from Annex I parties once again faced mostly officials from developing countries in the final friends group, which did very little for the effectiveness of that forum and contributed to the defection by the UK and US ministers into private, behind the scenes negotiations. Annex I delegates had reportedly warned the President that this would happen earlier in the session. Thirdly, the negotiating texts were still underdeveloped when ministers arrived (see also Chapter 11), and both officials and ministers from all parties expected that the officials would carry on working on these. The informal negotiating groups therefore fell between two camps. They were Co-Chaired by ministers who did not understand the texts and wanted a political discussion, yet almost all participants were officials – most ministers did not consider it their proper role to attend such day-to-day negotiations – who wanted to negotiate on the texts (Yamin and Depledge, 2004). The ministerial co-chairs were simply not equipped to chair such a textual discussion among officials, nor were the officials equipped to engage in a high-level political negotiation.

The experience of COP 6 highlighted another major challenge for the organization of the climate change negotiations, that is, the emergence of more and more issues that are both technical and political in nature, and therefore do not automatically fall within the purview of decision-making by either officials or ministers. During the post-Kyoto negotiations, these included, for example, eligibility rules for the flexibility mechanisms and a whole swathe of issues relating to the LULUCF sector, both highly complex matters with difficult technical implications. As Grubb and Yamin argued, 'fundamentally, the issues on the table at The Hague were too political for the technocrats to resolve, and too technical for the politicians to understand' (Grubb and Yamin, 2001, p269). Resolving such issues requires a more productive interplay between ministers and officials and their different strengths.

Such a productive interplay was better achieved at COP 6 (part II) and COP 7 (Yamin and Depledge, 2004), primarily by ensuring that ministers were only presented with the core issues requiring their high-level political input, and also through the use of shuttle diplomacy. At COP 6 (part II), for example, as noted in Chapter 11, officials working on the negotiating texts in the first few days of the Conference presented ministers arriving in the latter part of the week with a streamlined document clearly highlighting outstanding political issues and the options on the table (FCCC/CP/2001/CRP.8). Two ministers were then appointed to chair two of the informal arenas convened to discuss this document. Crucially, however, these arenas were designated as informal *consultations*, and not *groups*, so that the ministers concerned were not obliged to hold any actual meetings, but could just talk informally to the various coalitions and interested parties.

The President's text that followed shortly afterwards – the 'Core elements' paper – helped focus the minds of ministers even more. The use of shuttle diplomacy – where the President meets individually with representatives of each negotiating coalition – was then critical to overcoming problems with the disparity in participation between Annex I and non-Annex I parties. Although a friends group was convened, the real deal-making took place through shuttle diplomacy, where ministers and officials did not have to face one another directly (except, of course, for the President himself). This made for a more productive round of exchanges. Although ministers and officials did negotiate together in the final friends group on compliance, the dynamics were very different to the bad-tempered friends groups at COP 4 and COP 6 for two important reasons. Firstly, there were only three core issues requiring resolution, all of them highly political, so that the political/technical gap did not apply. Secondly, the developing countries negotiated as a bloc, through a South African spokesperson who was considered to be endowed with the authority of her minister, Valli Moosa, who had chaired negotiations on the issue but had now left the Conference. This, along with the determined yet poised personality of the individual negotiator herself, effectively bridged the gap in status between the interlocutors.

The experience at COP 7 similarly reveals the usefulness of shuttle diplomacy. Although a ministerial level friends group was convened, co-chaired by two ministers on behalf of the President, the real deal-making again took place through shuttle diplomacy by the facilitating ministers. In tandem with the official process of shuttle diplomacy, ministers were also very active in their traditional role of bringing the negotiations to fruition through unofficial talks behind the scenes.

Summary and concluding remarks

This chapter has shown how ministers can, and must, deliver high-level input into the climate change negotiations, unblocking issues where officials simply do not have the political mandate to compromise. This input, however, needs to be channelled judiciously in order to make productive use of the contrasting strengths of ministers relative to officials. The very different experiences of COP sessions in the climate change regime suggests that ministerial participation is most effective when channelled through shuttle diplomacy, or simply through guiding and instructing officials, wheeling and dealing behind the scenes, and bargaining on a few key issues at the last moment. The role of officials remains highly important right to the end of the negotiation process, especially when confronted with issues that cross the technical/political divide, as officials have more in-depth, longstanding knowledge of the issues and negotiating texts.

One of the major challenges in the climate change process is the greater level of ministerial participation – although not necessarily ministerial attendance – on the part of Annex I parties relative to non-Annex I parties. This does create problems in attempting to structure negotiating arenas based on ministerial participation. While increasing ministerial involvement on the part of non-Annex I parties is of course a long term goal, in the meantime, such strategies as shuttle diplomacy can avoid the rather unconstructive dynamics of most group meetings where Annex I ministers are confronted with non-Annex I officials.

Outside of the finales of major negotiating rounds, where ministers have been involved in the actual bargaining process, forums such as roundtables have been used to try to promote high-level political discussion and elicit ministerial input. This area is indeed one where the climate change regime has shown itself able to learn, abandoning old practices (such as the general debate), developing more innovative forums (the roundtables), and improving on these from session to session. The general consensus among interviewees, however, is that the climate change regime has not yet found ways of effectively harnessing the political leadership, skill and energy that ministers could yield. An alternative option could be to convene a high-level segment only once every two or three years, rather than at each annual COP, so that a stock-taking ministerial roundtable would have more meaning and purpose. Inviting that roundtable to actually negotiate a declaration to drive the future agenda of the COP could then be a useful means of helping to break the stale discussions that currently characterize much of the climate change negotiations.

Participation by Non-Governmental Organizations

The world is watching (the Climate Action Network, 1997).

Introduction

We now turn to examine channels in place for the participation in the climate change negotiations of non-governmental organizations (NGOs).[1] A huge variety of NGOs are active in the climate change regime. These encompass a very wide spectrum of differing objectives and shades of opinion, from environmental leaders to laggards, mirroring the spectrum of parties and indeed having greater extremes. NGOs carry out a variety of self-appointed roles in the negotiations. Some focus on raising awareness of climate change, others lobby for the interests of their own constituencies or the environment, an increasing number provide data, information and analysis, while still more are content with just observing proceedings of interest to their work. Openness to NGO input is widely seen as desirable in international environmental negotiations. As the climate change secretariat itself put it:

> The participation of NGOs is a fundamental element of the Convention process. It helps to bring transparency...facilitates inputs from geographically diverse sources and from a wide spectrum of expertise and perspective, improves popular understanding of the issues, and promotes accountability to the societies served (FCCC/SBI/2004/5, p4).

This chapter begins with an overview of rules for the admission of NGOs to the regime, before looking at the formal rules for their participation in the negotiation process. We then turn to more informal channels through which NGOs can input into the process, which, as we shall see, in fact tend to be more significant than the formal avenues.[2]

NGOs and the climate change regime

The climate change regime has always been relatively open to involvement from NGOs, with an expansive admission process and numerous channels for NGO inputs. This open approach was established from the outset by UNGA resolution 45/212, which launched negotiations on the Convention. The resolution explicitly 'invite[d] relevant non-governmental organizations to make contributions to the negotiating process' (paragraph 19). The resolution, however, also made clear the fundamental basis for NGO involvement in the climate change regime (akin to most other intergovernmental regimes) in confirming 'the understanding that these organizations *shall not have any negotiating role* during the process' (emphasis added).

The formal rules governing attendance and participation by NGOs that are set out in the Convention and rules of procedure are sparse, and mirror those in other regimes.[3] The climate change regime, however, has built on them to develop its own set of more detailed informal practices. In many ways, the development of NGO involvement has run in parallel with the three main negotiating rounds in the climate change regime.

Channels for participation granted to NGOs during the *Kyoto Protocol negotiations* were highly dependent on the discretion of the particular presiding officer, with the subsidiary bodies – including the AGBM and AG13 – adopting different approaches. The AG13, for example, was open and innovative in its elicitation of NGO input, whereas the AGBM was more conservative in its approach.

During this period, a debate was underway in the subsidiary bodies to consider different options for admitting and involving NGOs. This had its origins in a proposal put forward by New Zealand during the negotiations on the Convention to set up a 'business consultative mechanism' to enhance input by business and industry NGOs (BINGOs). The debate was interesting in itself for its participatory nature and openness to NGO inputs. A workshop was convened with NGO participation, individual NGOs were consulted on their views and ideas, and the secretariat put forward a set of proposals based on these inputs from NGOs. The secretariat's proposals, however, focused on improving existing practices rather than exploring new ones, as the SBSTA decided early on to circumscribe the debate in this way (see SBSTA 3 report, paragraph 50c). Despite the many innovative proposals that were put forward, the single outcome of this long process was thus the adoption of decision 18/CP.4 at COP 4, which established the default right of NGOs (and IGOs) to attend and observe meetings of contact groups (see below).

Other options were not pursued partly due to the absence of any common view among NGOs as a whole as to how their participation could be enhanced (op cit). This partly reflects the very different ways in which NGOs seek to make their views heard. The environmental NGOs (ENGOs), a more united group, tend to be vocal and visible in official forums (e.g. making statements, running side events, staging high profile, media-savvy demonstrations, issuing strong written position papers), while BINGOs tend to operate more on an individual basis behind the scenes, lobbying delegates and the organizers through chats in the corridors, private meetings and hospitality events.

It is also true that NGO participation during the Kyoto Protocol negotiations took place in a rather antagonistic context. Although NGO input was welcomed, the division of most NGOs into two camps – the 'green' ENGOs and the mostly 'grey' BINGOs – fostered an atmosphere of latent confrontation, especially among the more extreme sections of each camp. This followed through to parties, with developing countries in particular generally highly suspicious of NGOs. The approach of the business community as a whole was certainly more negative towards action to address climate change than it is today, with organizations such as the Global Climate Coalition and other business groups widely known to be lobbying hard within the US, and working together with obstructionist countries (notably OPEC) to block the negotiations.

The *post-Kyoto* period saw a more open attitude on the part of the presiding officers and secretariat to participation by, and inputs from, NGOs. This paralleled the emergence of a more constructive approach among much of the business community, as it began to recognize the business opportunities of a carbon-constrained world. As one BINGO interviewee acknowledged:

> over the last couple of years, we have gained more access, because of the acknowledgement that we have got something to offer... Things have improved since Kyoto primarily because industry has been much more constructive in the process.

Moves to secure the more constructive engagement of NGOs post-Kyoto were accompanied by the standardization of practices across the subsidiary bodies, notably the adoption of the above-mentioned decision 18/CP.4 on attendance at contact group meetings. Interestingly, however, NGOs have expanded their participation more through the gradual development of informal practices, than through the negotiation of formal rules. An important move, for example, has been establishing channels for NGOs to submit written views to the negotiation process, an innovation that is due as much to technical developments (the rise of the internet) allowing this to be done at low cost, than any lobbying on the part of NGOs (see also below).

The *post-Marrakesh* period has seen the issue of NGO involvement back on the formal agenda of the subsidiary bodies. This move was prompted by a conjunction of developments, notably the proliferation of workshops, where rules for invitation and participation by NGOs were deemed to require clarification and standardization. NGOs, especially from the business community, sought more extensive and established participation rights, and the US, sympathetic to their concerns, raised the issue in the SBI in 2002.[4] The ensuing debate prompted a more wide-ranging examination of prevailing practices governing NGO participation in the climate change process, including procedures for admittance, the constituency system, and channels for receiving inputs. The debate was ongoing at the time of writing, but appeared to be following the direction of the earlier debate on NGO participation during the period of the Kyoto Protocol negotiations, that is, to reaffirm, strengthen and codify existing practices, rather than develop any major new consultation mechanism.

Rules for admission

The right of NGOs to be represented at sessions of the climate change bodies is enshrined in the Convention and draft rules of procedure. FCCC Article 7.6 makes a distinction between two types of organizations. The first comprises UN bodies, that is, the UN, its specialized agencies and the International Atomic Energy Agency, along with states not party to the Convention, who 'may be represented at sessions of the COP as observers', that is, they are unconditionally admitted. The second consists of any other bodies or agencies, 'whether national or international, governmental or non-governmental', that is, both NGOs and IGOs. These organizations are subject to more detailed criteria for admittance, namely that they:

- should be qualified in matters covered by the Convention
- should inform the secretariat of their wish to be represented at a session of the COP as an observer and
- will be refused admittance if at least one third of the parties present object.

The rules of procedure repeat these rules and provide for their application also to sessions of the subsidiary bodies, while requiring the secretariat to notify observers of the date and venue of sessions to enable their representation (Rules 6–8).

A set of more detailed informal practices has become established within the secretariat to implement these formal rules on admission. These have been explained to the parties in documents prepared by the secretariat for the SBI, as part of the pre-Kyoto and post-Kyoto debates on NGO involvement (see FCCC/SBI/1997/14/Add.1, paragraph 3, and FCCC/SBI/2004/5, section II, respectively). As part of the pre-Kyoto debate, the SBI formally concluded that 'current arrangements for the accreditation of non-governmental organizations were satisfactory and that no change in the accreditation procedures was required' (SBI 8 report, paragraph 81a). As part of the post-Kyoto debate, parties once again took note of current accreditation practices, but, at the time of writing, decided to consider them further, due in part to concerns over the constituency system (see SBI 20 report, paragraph 100).

According to these practices, NGOs that wish to be admitted must fulfil three basic criteria to pass through a first screening by the secretariat:

1   They must be 'qualified in matters covered by the Convention' as required by Article 7.6.
2   Their governance structure must be independent of any national government.
3   They must confirm their non-profit, tax-exempt status.

In view of the all-encompassing nature of climate change, the constraint tends to be the demonstration of non-profit status, rather than of being qualified, usually interpreted as having an interest in climate change. The non-profit criteria requires businesses to group together in non-profit coalitions rather than represent themselves (Yamin and Depledge, 2004). Organizations that carry out functions on behalf of governments, or are funded by governments, are not necessarily excluded if their governing structure is itself independent.

The secretariat compiles a list of successful applicants based on these three criteria for the COP Bureau for its clearance. This is usually a formality, and it is rare, but not unheard of, for the Bureau to raise concerns. The list of cleared applicant NGOs (and IGOs) is then put to the COP for a formal decision on admission. Due to the large number of NGOs seeking admission, applications received in between sessions of the COP that have passed through the secretariat and Bureau may be presented to the next subsidiary body session for provisional admission, pending formal action by the COP (Yamin and Depledge, 2004). Once admitted, NGOs remain so. The exception is for certain very high-profile COP sessions (e.g. COP 3), where many NGOs (especially ones from the host country) were expected to attend only that session, and were therefore invited to reapply for admission to subsequent sessions. The admission of observers is very rarely challenged by the COP. As Wettestad (1999, p212) notes, the climate change regime thus seems to have adopted an 'overall inclusive model', with generous criteria for admission that have enabled a wide variety of organizations to attend regime body sessions, including those opposed to meaningful climate change mitigation action.[5]

The number of admitted NGOs and IGOs has risen steadily over the lifetime of the climate change regime. One hundred and seventy-seven NGOs and 20 IGOs were admitted at the time of COP 1. This rose to 298 NGOs and 31 IGOs by the time of COP 3, and to 619 NGOs and 50 IGOs by COP 9. The rise in admitted NGOs has not been accompanied by a parallel rise in the number of NGOs actually attending sessions (Yamin and Depledge, 2004). Less than half of admitted NGOs have been represented at COP sessions since COP 6 (and fewer at SBSTA/SBI sessions) compared with 79 per cent at COP 3 and 93 per cent at COP 1.

In terms of the number of individual participants, NGOs regularly constitute almost half the total number of delegates. This proportion tends to rise for high-profile finales of negotiating rounds. At COP 3 in Kyoto, for example, 64 per cent of participants were NGO representatives, with that figure reaching 56 per cent for COP 6 in The Hague (see FCCC/SBI/2004/5, paragraph 37).

An important dimension to NGO presence in the climate change regime is the great disparity in representation between NGOs from OECD countries on the one hand, and from developing countries and EITs on the other.[6] On the eve of COP 3, 91 per cent of NGOs admitted to the climate change regime had addresses in OECD countries (with 21 per cent in the US alone). The remainder hailed from non-Annex I parties, with only one NGO registered in an EIT (see FCCC/SBI/1997/14/Add.1, paragraph 13).[7] Although the situation has improved somewhat, more than 75% of NGOs are still based in Annex I parties, overwhelmingly from OECD countries. The disparity tends to be more striking in the BINGO constituency than the others. The main ENGO coalition, the Climate Action Network, usually fields representatives from its developing country and EIT chapters (e.g. CAN-South Asia, CAN-

Africa, CAN-Central and Eastern Europe). The Indigenous People's Organizations (IPOs), although small in number, are almost entirely developing country based, while the Research-oriented and Independent NGOs (RINGOs) also include a handful of developing country research institutes and universities. The BINGOs, however, hail almost exclusively from OECD countries, with only a small handful of developing country or EIT delegates occasionally present.

While the climate change regime extends funding to assist government delegates from developing countries and EITs to attend sessions of the regime bodies, there are no similar measures in place to support and promote the attendance of NGOs from those countries. NGOs, especially ENGOs, have long called for such funding (Yamin and Depledge, 2004), and it has featured in secretariat proposals considered in both the pre-Kyoto (see FCCC/SBI/1997/14/Add.1) and post-Kyoto (see FCCC/SBI/2004/5) debates. Given that the state of the participation fund for eligible *parties* remains precarious, the likelihood of parties funding NGO participation is slim.

NGO constituencies

An informal practice has emerged over time within the secretariat of recognizing *constituencies* of NGOs (Yamin and Depledge, 2004). Five of these are currently in place:

* Environmental NGOs (ENGOs)
* Business and Industry NGOs (BINGOs)
* Local Government and Municipal Authorities (LGMAs)
* Indigenous People's Organizations (IPOs) and
* Research-oriented and Independent NGOs (RINGOs).

The constituency system has developed on a bottom-up, pragmatic basis (Yamin and Depledge, 2004). ENGOs and BINGOs are the most longstanding constituencies, dating back from the first session of the Intergovernmental Negotiating Committee (INC) that negotiated the Convention, and are commonly represented in almost all environmental negotiations. They organized themselves into umbrella groupings early on in the climate change process, in much the same way as parties have organized themselves into negotiating coalitions. These are still the two most active constituencies. The LGMA constituency was added at COP 1, following high levels of activity and lobbying by the constituency at that session, although it has been less active since. The IPOs emerged as a new voice in the climate change regime post-Kyoto, and were recognized as a constituency at COP 7 in 2001. Despite being represented by only a few individuals at negotiating sessions, they have been vociferous in calling for special recognition on account of the situation of their members, whose livelihoods they argue will be directly affected by COP decisions (e.g. on the CDM). They have pursued these concerns in other parts of the international arena. A Permanent Forum on Indigenous Issues, for example, was established as an advisory body to the UN's Economic and Social

Council (ECOSOC) in 2000, and has made recommendations to increase involvement of IPOs in the climate change regime, including by establishing an inter-sessional working group on indigenous peoples and related issues (see FCCC/SBI/2004/5). The RINGOs are the latest constituency to be added to the list, achieving recognition at COP 9 in 2003. This constituency consists of universities and other independent research institutes, who decided to group together in response to a sense that they were missing out from the better access channels afforded to the existing constituencies.

The five constituencies are not exclusive. There are, for example, NGOs that would describe themselves as 'environmental' who are not part of the ENGOs, while the newly formed RINGOs currently include only a minority of research-focused NGOs active in the regime. Similarly, there are many NGOs – trade unions, women, parliamentarians, faith-based groups and others – who have not formed their own constituencies (Yamin and Depledge, 2004).

Although it is only an informal practice, without roots in any formal rules, the constituency system is now well established. It is used by the secretariat to organize participation in the negotiation process, including speaking slots at the high level segment, invitations to workshops, channelling of written inputs (see below), organizing meetings with the presiding officers and other dignitaries, and consultation on procedural issues (e.g. the code of conduct discussed below). The secretariat liaises with the designated focal point of each constituency, which then organizes the required input among the constituency members. Individual NGOs that are not in any constituency may also participate in the negotiation process and liaise with the secretariat in the same way. Nonetheless, several members of the newly-formed RINGO constituency reported in their interviews that access had improved since the establishment of their constituency.

The constituency system therefore helps the secretariat, as well as NGOs themselves, who can pool their resources. The NGO constituencies tend to meet daily during negotiating sessions to coordinate their work and exchange information. The constituency system is thus a pragmatic, innovative and simple response to the complexity of the climate change regime, permitting more meaningful and efficient participation by the 600+ NGOs in the regime than would otherwise be possible. The system is not, of course, unique in the international environmental arena. Agenda 21,[8] for example, recognizes nine major groups.

In line with the informality of the constituency system, no formal procedures exist for setting up a constituency. The process through which the IPO and RINGO constituencies were formed, however, established certain precedents. One of these is that the secretariat has served as the gatekeeper, arbitrating on whether, and how, new groups of NGOs can become a constituency. The formation of the RINGO constituency, for example, involved several exchanges of correspondence between NGO representatives and the Executive Secretary. According to the secretariat website, NGOs wishing to set up a constituency should now fulfil the following criteria:

- include a critical mass (unspecified) of members
- provide a focal point for liaison with the secretariat

- participate regularly at negotiating sessions and provide common inputs (e.g. statements) and
- have channels in place for information exchange among members.

The actual process for considering a request based on these criteria is unclear. In the case of both the IPOs and RINGOs, the COP Bureau gave eventual approval to their requests, so that they were recognized as constituencies in 2001 and 2003 respectively. However, as decisions of the COP Bureau and minutes of its meetings are not made available (see Chapter 5), there is no official accessible record of how this decision was made.

The constituency system is thus a very good example of an informal established practice that emerged over time, in this case through the combined actions of the secretariat and NGOs, in the absence of any rules, guidance or endorsement from the parties through the COP or SBI. The absence of any set procedures for establishing a constituency perhaps helps explain why, between COP 1 and COP 6, there were no requests by any group of NGOs to form a constituency, despite the wide variety of active NGOs other than the existing ENGOs, BINGOs and LGMAs. Private conversations with research NGOs suggest that the possibility of establishing a new constituency had simply not been considered as an option. The option only came into the open following the request for a more effective voice by indigenous peoples, with the secretariat and the COP Bureau seeing the formation of a new constituency as an easy (although for the IPOs unsatisfactory) means of addressing their concerns. The RINGOs then followed suit, once the possibility of doing so had been opened up.

The experience of the constituency system points to one disadvantage of unwritten informal practices, even if they are well-established; because the practices are not necessarily clear to all, or are interpreted differently, ability to make use of them is rather arbitrary. Another disadvantage is that when, perhaps inevitably, some controversy emerges, the established practice is much more vulnerable to challenge. While a formal rule can be defended on the grounds that it was formally adopted through accepted procedures, an established practice can only appeal to precedent or logic. The constituency system and its use of focal points, for example, were challenged by the US in the SBI in the post-Marrakesh period. Revealingly, the US stated 'To our knowledge, there are no "recognized constituency groups" in the Convention, nor have *we the parties* ever identified any such groups' (see FCCC/SBI/2002/MISC.8, emphasis added).[9] At the time of writing, the constituency system was being subject to review by the SBI as part of the wider debate on NGO participation.

Channels for participation

The code of conduct

A noteworthy development in the post-Marrakesh period has been the devising of a code of conduct to guide participation by NGOs in the climate change negotiations. This code of conduct – officially known as 'Guidelines for the participation of representatives of non-governmental organizations at meetings of the bodies of the United Nations Framework Convention on Climate Change'[10] – was prepared by the secretariat in consultation with the NGO constituencies. It responded to a small number of instances where individuals attending under the badge of environmental NGOs were responsible for disruption to the negotiation process and the harassment of some delegates (Yamin and Depledge, 2004).

The code of conduct is not an official document, and has not been the subject of negotiation in the subsidiary bodies. Rather, it represents an attempt by the secretariat to set out mutual expectations for the behaviour of NGOs at negotiating sessions, 'reflecting current practice' and based on guidelines 'governing NGO participation' elsewhere 'in the UN system'. Most of its provisions are self-evident – no harassment or restriction of movement, no unauthorized demonstrations within UNFCCC venues, no disrespectful treatment of country flags, maintenance of respect for participants' 'social, cultural, religious and other beliefs' – and, if they constrain the legitimate activities of reputable NGOs, do so only marginally. Nevertheless, the code of conduct, despite its informal nature, was not universally welcomed, with some NGOs viewing it as a sign of mistrust of their work, and possibly also a slippery slope that could be used to impose greater restrictions in the future.

Observing

The most basic form of participation by NGOs in the climate change regime is through observation of the negotiations. This forms the basis for maintaining the transparency of the process, as well as the accountability of national delegates to their populations back home. As such, observing is not necessarily just a passive act. By observing the negotiations, NGOs can obtain the information they need to follow the negotiation process, monitor the positions of governments, develop their own stance, and report back to their members.

The flipside of the transparency that NGO presence provides, however, is the concern that it can discourage parties from showing flexibility or exploring possible trade-offs. The absence of scrutiny from NGOs and the media is indeed commonly viewed as pivotal to encouraging parties to speak more freely. Explaining his decision to close the AGBM non-groups to observers during the Kyoto Protocol negotiations, Estrada said in his interview 'I do think that negotiations have to be private...It doesn't make sense to have people negotiating with NGOs [present]... not because there is something to hide, but it is difficult for people to modify positions when they are being watched'. Many party interviewees agreed that they could not envisage serious negotiations in the presence of NGOs. One interviewee remarked:

> You cannot have a contact group that is open to discuss sensitive issues...of course if they do it, they [the negotiators] are going to maintain their position, because they know they [the NGOs] are there...they will not move a centimetre. But if you are in smaller group...talking to people who can understand you, maybe

because they are civil servants and have similar problems, then you can open up, and you can say what you couldn't say.

This helps explain why the rights of NGOs to observe proceedings diminish with the greater informality of the arena (see Chapter 9).

Plenary

NGOs have always been allowed to observe COP and subsidiary body plenary meetings.[11] Squabbles sometimes arise over seating space, but these are logistical in nature and typically soon resolved.

The only major exception to this was the established practice in place during the Kyoto Protocol negotiations whereby NGO representatives were, as a general rule, prohibited from coming onto the main negotiating floor during plenary meetings. This practice dated back to an incident just before COP 1 in 1995. At the 11th (and final) session of the INC, objections were raised to the presence of fossil fuel BINGO lobbyists on the negotiating floor, whose advice to OPEC states appeared to forestall an emerging consensus on the rules of procedure. Despite protests by NGOs at their subsequent exclusion from the negotiating floor, the COP Bureau maintained the general practice, but allowed the Chairs of each subsidiary body to exercise discretion in granting access. The SBSTA Chair, for example, took a more relaxed attitude, providing a seat for each NGO constituency on the meeting room floor (e.g. see SBSTA 6 report). Chair Estrada, however, who had been chairing the INC 11 meeting at which the above-mentioned incident occurred, applied the practice strictly in the AGBM. The established practice gradually lost its relevance when the secretariat moved its headquarters from Geneva to Bonn, whose conference venues do not have such clearly separated spaces for NGO seating as the UN building in Geneva. In addition, the advent of mobile phone technology now allows NGOs to communicate with delegates without coming onto the floor. One interviewee recalled, 'at INC 11, when the Kuwaiti delegate got up to go to the toilet, a load of delegates followed him to ensure he didn't talk to anyone! Now with cell phones it's all different. Technology has affected the process.' It is not unusual for sharp-eyed onlookers to witness a mobile telephone conversation between a country delegate on the negotiating floor and an NGO representative seated in the gallery, at the conclusion of which the country delegate will raise his flag to make an intervention, quite clearly on advice of that NGO.

Informal groups

The adoption of decision 18/CP.4 means that NGOs are now routinely permitted to observe proceedings in open-ended contact groups, unless at least one third of parties object. This marked an important change from the Kyoto Protocol negotiations, where NGOs were typically not given access to informal groups. As noted above, for example, Chair Estrada closed the informal non-groups he established from AGBM 6 in March 1997 onwards. This aroused strong objections from NGOs, illustrating the importance that most attach to observing. The journal ECO reported:

> NGOs are very disappointed by the Chairman's decision to exclude us from important negotiating sessions. Simply through the establishment of a new category – non-groups – at yesterday's meeting, the rules of transparency and public participation no longer apply (ECO, 1997a).

Estrada's reasons for closing the non-groups were founded not only on the assumption that parties would be more reticent to negotiate in the presence of NGOs, but also on genuine concern that the presence of obstructionist NGOs might hamper the negotiations. The situation changed at the COP 3 finale. From the second week, NGOs were admitted to observe proceedings at all meetings chaired by Estrada, although the other informal groups remained closed. The last night of negotiations was completely open to all observers, including television cameras, when Estrada actively sought out public scrutiny to place pressure on parties to reach agreement.

This case illustrates well the contrasting roles of NGOs, and how these can be more or less valued at different points in the process. When he wanted to encourage parties to intensify their negotiations in the non-groups, Chair Estrada sought to remove the scrutiny and transparency that comes with NGO presence in order to give parties the necessary privacy to engage in bargaining. However, for the final deal-making on the last night of COP 3, he sought to make use of the transparency and scrutiny provided by NGOs that he had previously curtailed, as a strategy to place the strongest possible pressure on parties to compromise and not block agreement.

Even with the adoption of decision 18/CP.4, presiding officers of contact groups may still close the group to observers at any time. This has often been done when contact groups have begun to engage in serious bargaining in the latter stages of negotiations. Decision 18/CP.4 also refers specifically to 'open-ended contact groups'. However, as noted in Chapter 9, the informal groups convened in the climate change regime span a wide spectrum and have been known by a variety of different names, especially at negotiation finales. Participation by NGOs in any forum not known as a 'contact group', therefore, is essentially at the discretion of its presiding officer (Yamin and Depledge, 2004). The ministerial 'cluster groups' and negotiating groups at COP 6, COP 6 (part II) and COP 7, for example, were all closed to NGOs. Moreover, there is no compulsion on Chairs of informal *consultations* to involve NGOs.

Access to documentation

Meaningful observation and understanding of proceedings requires access to the documentation under discussion. NGOs have always had access to all documentation distributed in the room in which they are present, although often in limited supply and later than the parties. Indeed, all *official* UNFCCC documents are made available to NGOs, and also to the public at large, through the UNFCCC website. There is no such thing as a restricted document in the climate change regime. This does not, however, apply to all *unofficial* documentation, namely, the non-papers (for an explanation of the main types of official and semi-official documentation, see Table 11.1). Where these are distributed in a meeting in which NGOs are present, or made generally available, NGOs will also receive a copy, subject to the above provisos. Where the meeting is closed to observers, NGOs will need to lobby delegates or the organizers outside the room to obtain a copy, and indeed will often be successful in doing so.

Making statements

Making statements in plenary meetings has long been an established channel for NGOs to input into the climate change negotiations. At its very first session, for example, the INC gave effect to the explicit call in UNGA resolution 45/212 for NGOs to 'make contributions' to the negotiations on the Convention (see above) by inviting 'two observers representing different groups of non-governmental organizations [in effect, the BINGOs and ENGOs] ... to speak at the end of the general debate'.

Statements can enhance transparency by providing a channel for NGOs to give feedback to government delegates on developments in the negotiations, as well as to formally communicate information and ideas on possible innovative solutions to the problems facing the negotiations. NGOs are granted several speaking slots during COP sessions under the standing COP agenda item 'Statements by NGOs'. These speaking slots are allocated by the secretariat to the constituencies and then according to demand, with the aim of securing a representative range of speakers. As illustrated in Table 14.1, a wide spectrum of different NGO groups are typically represented, including, for example, trade unions and faith groups, who very rarely speak at other times in the process. The two or three statements by ENGOs and BINGOs reflect the efforts made by those constituencies to field locally-based members, as well as representatives of their international chapters. In the case of the BINGOs, it also reflects differences in perspective among their members, who would therefore be unable to agree on a common statement.

Table 14.1 *Statements made by NGOs at COP sessions*

COP	ENGO	BINGO	LGMA	IPO	RINGO	Parliament	Youth	Faith	Labour	Other
1	√	√	√		√√	√				
2	√√	√√	√√					√		
3	√√	√√√	√			√	√	√	√√	√¹
4	√√√	√√√	√√			√		√	√	√
5	√√√²	√√√	√√					√		√³
6	√√√²	√√√	√	√		√	√	√	√	√³
7	√√	√√√		√		√	√	√	√	
8	√	√√√	√	√	√		√	√	√	
9	√√√	√√	√	√	√			√	√	

Note: 1 Scientists for global responsibility.
Note: 2 Internation Union for the Conservation of Nature.
Note: 3 European Landowners' Organisation.

These statements, however, are almost always general and highly rhetorical in nature, akin to the equivalent statements made by ministers and other heads of delegation in the general debate (see Chapter 13). The statements do serve an important purpose in allowing NGOs to put their views on record and thereby influence the general thrust of the international climate change agenda. One ENGO interviewee, for example, offered the following perspective on the statement by the Climate Action Network at COP 3:

> To bring attention to what's really happening can be a really powerful thing to do. It was in the second week, and everything was bogged down, and it was a reminder, a pep talk, of why are we here... in the lead up to Kyoto, the plenary statements were the only opportunity [to participate] so they were important.

It is doubtful, however, whether such rhetorical NGO statements impact on the actual negotiation process in any practical way at all. As one interviewee put it, 'The interventions by NGOs are, 95 per cent of the time, predictable, so there are no surprises whatsoever, make that 99 per cent of the time, predictable'.

NGOs are also permitted to make statements in the subsidiary bodies, usually on the basis of demand and at the discretion of the presiding officer. The extent to which NGOs have been granted speaking slots, and indeed have asked for them or taken them up, has varied considerably in the climate change regime (Yamin and Depledge, 2004).

During the Kyoto Protocol negotiations, Chair Estrada typically allocated one time slot per negotiating session for each of the three (at the time) constituencies to make a general statement. These statements were not afforded a prominent place, usually being scheduled at the very end or beginning of a meeting, when many delegates were drifting in and out and not fully attentive. It was only at the last three sessions of the AGBM that all three constituencies availed themselves of the opportunity to speak, and only the ENGOs spoke at every AGBM session. The absence of statements by BINGOs in the early sessions can be attributed to the differences of opinion between different factions, which meant that the constituency as a whole was unable to agree a common statement. A key drawback to the practice for NGO interventions adopted by the AGBM was that, because only one statement was allowed per constituency per session (eventually two for BINGOs), that statement was likely to be of a general nature in order to cover all the issues on the table.

The SBSTA and AG13 Chairs during the pre-Kyoto period took a different approach, allowing NGO statements on *specific agenda items* on behalf of constituencies. These statements were taken after those by parties, but nevertheless as part of the main debate. This permitted more targeted interventions, including specific recommendations and responses to developments in the negotiation process. NGOs, however, seldom took up the opportunity afforded to them, with only a few NGO statements recorded in the SBSTA, AG13 or indeed the SBI, during the pre-Kyoto period. The AGBM, therefore, with its more formal single speaking slot, in fact generated more input than the more open approaches of the SBSTA and AG13.

The approach of allowing interventions on specific agenda items by NGO constituencies has continued in both the SBSTA and SBI in the post-Kyoto period. The start of a new exploratory stage on the novel issues raised by the Kyoto Protocol, along with the more open attitude of the SBSTA and SBI Chairs, led to a sudden upsurge in demand by NGOs to speak in the early post-Kyoto negotiations. This was especially so in the SBSTA, which is responsible for issues traditionally of particular concern to ENGOs, such as LULUCF, bunker fuels and the relationship with the ozone layer. Between two and four NGO statements were made at each session between SBSTA 8 (early 1998) and 11 (late 1999). As in the AGBM process, the overwhelming majority of statements in the subsidiary bodies were made by ENGOs, reflecting their more united positions and the greater value that they attach to participation in official arenas.

However, as negotiations moved on from the exploratory stage to engage in more detailed bargaining as the COP 6 deadline approached, the relevance of plenary debates relative to more informal arenas declined, and therefore the value placed on plenary statements similarly decreased. This decline has continued post-Marrakesh as the focus of the process has shifted towards implementation. Only one NGO statement was made in the SBSTA between SBSTA 12 (mid 2000) and SBSTA 19 (late 2003), and none at all in the SBI. NGOs did, however, make targeted interventions on specific agenda items at COP 9 in 2003, the first time that they had done so in a COP plenary. The ENGOs, BINGOs and IPOs all intervened in the debate on the CDM-Executive Board report, while the ENGOs and LGMAs spoke on the topic of the review of Annex I party commitments. The fact that NGOs did ask to speak in the highly formal COP plenary on these two key items illustrates that, where the plenary debate is seen to be of relevance, NGOs will seek to intervene.

At the same time, in the post-Kyoto period, more and more effective channels have been opened up or enhanced for NGO participation, such as the opportunity to participate in contact groups, more structured and regular meetings with the presiding officers, attendance at workshops and other complementary forums, and submission of written inputs for electronic circulation on the secretariat website (see below). These channels are generally seen as more efficient and effective in putting forward NGO viewpoints than the traditional channel of plenary statements. One of the main obstacles to greater NGO input in plenary debates is the requirement that NGOs speak on behalf of a broader international constituency (but not necessarily the five recognized ones). This means that NGO delegates observing a meeting cannot spontaneously ask for the floor to respond to developments in the debate, as any statement would first need to be agreed within the constituency as a whole. This can involve a lot of work for the constituencies. In particular, it helps explain why BINGOs, whose views are much more diverse than the ENGOs, have made so few plenary statements. Moreover, the practice whereby NGO statements are taken at the close of the debate means that NGO delegates cannot contribute at whatever point they feel is most appropriate, perhaps missing the chance to make the most impact.

NGOs are occasionally permitted to speak during contact group meetings, at the discretion of the Chair. The same restrictions, however, apply as during plenary meetings, and NGOs do not participate in the actual bargaining process (see FCCC/SBI/2004/5, paragraph 25).

Participation in workshops and other complementary forums

A more effective means of eliciting contributions from NGOs than plenary statements has been their participation in complementary forums, including the informal roundtables convened during the Kyoto Protocol negotiations, and the post-Kyoto workshops (on complementary forums, see Chapter 10).

During the AGBM roundtables, which were open to all participants, NGO representatives were allowed to speak more than once, and outside the constituency structure, providing an important opportunity for them to communicate ideas to delegates in a forum that was precisely aimed at such an open exchange (see also Chapter 10).

The post-Kyoto workshops have similarly seen strong NGO participation, although this has varied depending on the nature of the workshop and the level of interest by NGOs. Disquiet among NGOs in the early post-Kyoto period over whether, how, and in what numbers they were invited to workshops was partly what prompted the wider discussion of NGO participation by the SBI in the post-Marrakesh period. Invitations to NGOs are now issued by the secretariat through constituency focal points, who are canvassed in advance to assess the level of interest in participating in the workshop (Yamin and Depledge, 2004). The secretariat reports that almost all of the 14 workshops held in 2003 had representation from NGOs, although not from all constituencies (see FCCC/SBI/2004/5, section F). Once again, the ENGOs were relatively well represented, along with the RINGOs. Importantly, the number of invitations generally exceeds the number of eventual participants, suggesting that formal channels for securing participation by NGOs are adequate. The main constraint on attendance at workshops by NGOs is rather lack of time and resources. This is borne out by a BINGO interviewee, who confirmed that the real constraint facing his organization was finding colleagues prepared to spare the time to attend workshops, and by an ENGO interviewee, who blamed low levels of attendance on lack of financing.

Actual participation by NGOs at workshops is at the discretion of the Chair, but typically NGOs are permitted to intervene more frequently – not just once and not just after parties – and as individual organizations, rather than constituencies. This makes for a much freer and more meaningful process of inputting into the discussions. NGO representatives regularly also act as resource persons at workshops, making presentations or otherwise providing technical information. Overall, although NGO representation at workshops is still low in numbers relative to parties, these complementary forums do tend to provide a useful channel for NGOs into the process, albeit in terms of providing information and technical analysis, rather than lobbying for particular positions.

Consultations with the organizers

An important means for NGOs to convey their views is through private meetings with the presiding officers and secretariat, mirroring the bilateral consultations that are common between parties and these organizers of the negotiation process. Since 1991, the Executive Secretary has met privately with each NGO constituency at each negotiating session. Former Executive Secretary Michael Zammit Cutajar reported that such meetings usually cover a wide spectrum of issues, from the substantive views of NGOs on issues under discussion, to procedures for their participation and logistical facilities offered to them. Interviews indicated that NGOs appreciate these meetings, enabling them to communicate their views and feel more involved with the process. One BINGO interviewee stressed how valuable he had found meetings with the new Executive Secretary Joke Waller Hunter, allowing his constituency to gauge better her character and the nature of her leadership, enabling it to better direct its work in the climate change regime.

COP Presidents and subsidiary body Chairs similarly meet privately with each NGO constituency. Although Chair Estrada did not consult with NGOs on a regular basis during the Kyoto Protocol negotiations, he did so on specific occasions, for example, to introduce his Chair's text or to hear procedural grievances (e.g. exclusion from the non-groups). General meetings between the subsidiary body Chairs and NGO constituencies became established practice in the post-Kyoto period, prompted partly by the SBSTA Chair's strong enthusiasm for engaging NGOs. Again, these meetings typically cover a range of issues. ENGOs usually focus more strongly on conveying their substantive positions than do the BINGOs, who generally concentrate on procedural matters, reflecting the greater diversity of views within their constituency.

Presiding officers have sometimes used private meetings with NGOs as a means of compensating for their exclusion from negotiating arenas. During the Kyoto Protocol negotiations, for example, Chair Estrada organized daily 'briefings' exclusive to NGOs to provide them with information and the opportunity to ask questions on the closed non-groups.

The extent to which meetings between the presiding officers and NGOs actually feed into a substantive impact on the negotiation process will largely depend on the goodwill of the presiding officer. The frequent interaction between NGOs and the organizers of the negotiation process, however, has contributed to the gradual trend in strengthening their engagement in the regime.

Submission of written inputs

The COP and subsidiary bodies regularly invite *parties* to make written submissions and proposals on issues under negotiation, which are then compiled, printed and circulated as official UNFCCC 'MISC' documents (see Chapter 11). Traditionally, however, NGOs have not been invited to make submissions in this way, and any unsolicited NGO submissions have not been included in the documentation of the regime (Yamin and Depledge, 2004). This reflects the view that NGOs are not involved in the actual negotiations, along with logistical concerns at the potential costs of reproducing and circulating NGO submissions.

The growing sophistication of internet technology, however, means that documents can now be placed on the internet at low cost, and do not have to be reproduced in hard copy to be made widely available. This has removed a key barrier to the acceptance by the climate change regime of written submissions by NGOs. The first use of the internet for this purpose occurred for inputs on ways and means of limiting emissions of hydrofluorocarbons and perfluorocarbons. On this issue, COP 4 explicitly called for submissions from IGOs and NGOs (decision 13/CP.4), in addition to parties, in the knowledge that much technical expertise on this issue is located within the business and industry community. Submissions from IGOs and NGOs were simply placed on the secretariat website, along with submissions from parties, which were also reproduced in MISC documents. COP 5 invited 'each Party to give consideration' to this information, and the IPCC to take it into account in the preparation of the TAR (decision 17/CP.5). In this case, therefore, submissions from NGOs and IGOs were given almost equal prominence to those from parties.

Building on the potential of internet technology, the secretariat has created a new form of semi-official UNFCCC document, the 'web-only' document, that is made available only via the secretariat website and not in hard copy (Yamin and Depledge, 2004). The web-only document has become a very useful vehicle to enable the circulation of written inputs submitted by NGOs (see also Chapter 11). This development has, in turn, enabled the subsidiary bodies to request submissions from NGOs and IGOs as well as parties, without fear of the procedural/logistical implications. SBI 16, for example, invited parties, *NGOs and IGOs* to provide information on their experience regarding the effectiveness of the Global Environment Facility (GEF) as the financial mechanism of the Convention. The submission presented in response by the Climate Action Network was duly issued in a web-only document (FCCC/WEB/2002/6), and referenced as a document for consideration by the SBI in its subsequent discussion of the topic. The same process was followed for the high-profile issue of the inclusion of LULUCF projects under the CDM, where parties *and other organizations* were invited to submit their views as part of the workplan on this issue, resulting in submissions from three ENGOs and one RINGO (FCCC/WEB/2002/12). Interestingly, the submissions were accepted from individual NGOs, rather than constituencies as a whole. In two other cases, the SBSTA accepted unsolicited submissions from NGOs – in both cases ENGOs – and issued them as web-only documents (FCCC/WEB/2002/13 and FCCC/WEB/2002/14). In all these cases, NGO submissions were clearly separated from those of parties, making clear the higher status of the latter. The practice of circulating NGO submissions in web-only documents has now been formally endorsed by the SBI (see I SBI 20 report, paragraph 104), two years after the secretariat launched the initiative.

The official publication of NGO views in this way is an important procedural development. However, there has not been much uptake of the opportunity afforded to NGOs to make submissions in this way. Almost all the submissions have been from ENGOs, with none at all from BINGOs; this suggests that BINGOs place less value on this means of influencing the intergovernmental process, preferring to focus on lobbying individual national governments. Even among ENGOs, there has been no attempt to input on all issues where submissions have been invited from parties, and indeed no NGO submissions were received in 2003. This again suggests that NGOs as a whole prefer to devote resources to informal means of influencing the process, rather than engaging in formal channels.

Side events and exhibits

In addition to the formal vehicles for participation, a tradition has emerged in the climate change regime of holding *side events* on the margins of the official meetings (FCCC, 2000). These side events, organized mostly but not exclusively by NGOs and IGOs, consist of a diversity of activities, including seminars, workshops, presentations, panel discussions and debates. Many NGO delegates attend negotiating sessions primarily to take part in the jamboree of side events. The secretariat allocates a fixed number of time slots for side events in advance of the session, and the events are then publicized in the official daily programme of meetings (Yamin and Depledge, 2004).

The number of side events organized by parties, in particular, has risen over time, as they have increasingly taken advantage of the opportunity to present aspects of their climate change actions and policies in an informal, non-political way. Even the subsidiary bodies have recently chosen to make use of the side event environment as part of the formal consideration of an agenda item. Such cases are discussed more fully in Chapter 10.

The growing popularity of side events reflects several trends at work in the climate change regime. Firstly, it reflects the great, sustained and rising interest in the climate change regime on the part of civil society along with the diversity of topics and sectors involved in tackling climate change, which together generate a plethora of organizations and individuals wishing to publicize their work, or to make their ideas, proposals and perspectives heard. Table 14.2 below illustrates the type of activities taking place as side events, and their organizers.

A second, very important spur to the growth of side events is the unrelenting politicization of the formal climate change negotiations, which forecloses many avenues of debate on the grounds of political sensitivity. So many subjects have become taboo that side events have emerged as an alternative safe outlet for open discussion and debate on these topics (see also Chapter 10). The best example is that of future commitments for all parties; in the formal negotiations, the item that would provide a home for such a discussion has not even made it onto the formal agenda, and instead has been held in acrimonious abeyance for the past five years. It is no exaggeration to say that the mere mention of future commitments or negotiations on these is taboo in the formal regime. However, recent negotiating sessions have seen a wealth of side events on this very topic, usually with presentations and in-depth debate involving both developed and developing country NGO delegates. As one interviewee argued 'side events, particularly around the shaping of the debate post-Kyoto, are critical, as the system itself isn't capable of shaping that kind of debate, and there's a very rich debate going on'. Other topics dealt with in this way include presentations and discussion of specific climate change policies and emitting sectors (e.g. transport, electricity generation). A partnership of BINGOs, ENGOs and RINGOs, for example, organized a high-level event on options for tackling transport emissions at COP 9. Such a debate, despite its critical relevance to tackling climate change, could never take place within the confines of the negotiation process, with the debate on policies and measures indeed stalled in the SBSTA. As ENB (2003b) reported, side events 'forged ahead into unchartered waters, exploring those issues that have proved simply too hot to be handled by the COP'.

Several commentators pointed to COP 9 as a turning point in this respect, where the contrast between the sluggishness of the formal negotiations and the energy of the side events was thrown into focus. Several interviewees at COP

Table 14.2 *Extract from the daily programme of meetings for Wednesday 10 December 2003 at COP 9*

Event title	Organizer
European greenhouse gas budgets of the biosphere	European community
Windpower and climate change Adaptation to climate change risks in Small Island States	Greenpeace International Delegation of Samoa with CARICOM
Perspectives for the further development of the Kyoto Protocol Emissions trading: the financial sector	German Advisory Council on Global Change UNEP
Linking climate responses and development planning	OECD
GHG reporting guidance: comparison of two approaches	Natural Resources Defense Council
Choices and challenges: How Arkhangelsk pulp and paper mill (Russia) is reducing emissions while achieving business growth	Environment Defense
Climate change in the Arctic: human rights of the Inuit interconnected with the world	Center for International Environmental Law
Rethinking tropical deforestation and the Kyoto Protocol	Delegation of Italy

Source: FCCC/CP/2003/OD/9

9 echoed the view of an ENGO respondent who said: 'here, and also in Delhi [COP 8], the side events were better than the meetings'. Dessai et al similarly quoted some observers as saying that 'the most valuable outcomes of COP 9 occurred during the side-events' (Dessai et al, 2004). This has not always been the case. As ENB (2003b) commented:

> in previous years, some delegates . . . considered the side events to be an opportunity for free food and a quick nap during the lunch period. But at COP-9, the side events also provided considerable food for thought . . . while the official negotiations crawled along at a snail's pace.

An interesting aspect of side events is that they are overwhelmingly focused on positive responses to climate change, often far ahead of the formal negotiations among parties in terms of the strength and innovativeness of action proposed or indeed being undertaken. ENB (2003a) reports that 'while diplomats were often left agreeing on the lowest common denominator ... almost all [the over 100 side events] were focused on the highest common denominator'. The massive majority of side events were extremely optimistic about the possibility of implementing meaningful action to tackle climate change in an effective manner. This is particularly significant given that most side events were not organized by ENGOs, but rather by research based NGOs. There is no doubt that the most interesting, constructive and potentially integrative solutions to the dilemmas of the climate change negotiations are proposed, developed and discussed in the more problem-solving atmosphere of the side events rather than the official arenas.

Side events, however, do indeed remain 'on the side' rather than central to the negotiation process.[12] Despite the larger than usual number of party negotiators attending side events at COP 9, most participants at side events are traditionally other NGOs. The side events do indeed provide a very important forum for NGOs to network, as well as to communicate and debate their views and ideas on how the negotiations should proceed. As one NGO interviewee commented:

> special events are extremely useful, to present our own research input or to get latest news of research being done by others . . . due to the proliferation of academic NGOs in the meetings, you get a lot of papers giving the latest scientific input, grey literature, which would not be published years from now.

Although NGO side events rarely impact directly and immediately on the formal negotiations, they do have a very important contribution to make in terms of stimulating and encouraging the broader long-term flow of ideas on tackling climate change, elements of which can eventually filter through to the actors on the negotiation stage. Another important contribution of side events – not only from NGOs, but also from IGOs and parties – is to showcase the implementation of projects and programmes related to climate change on the ground, which often contrast with the greater hesitation of parties in the formal negotiations to commit to strong climate change action for fear of economic harm.

One party interviewee suggested that side events are becoming increasingly important in triggering debate (if not shaping positions) even among parties: '[government] delegates are attending. There's quite a lot of discussion that flows from that. It permeates through.' Indeed, in many cases, party delegates attending side events appeared to feel relieved at being able to broach topics in a wholly informal, exploratory manner, without fear of censure for political incorrectness.

Nevertheless, links between side events and the formal negotiations remain weak, so that it is difficult for debates at side events to have any short-term impact on the negotiation process. No formal record is kept of discussions, and there are no channels for reporting results or insights to the subsidiary bodies. However, reflecting the increasing importance attached to side events, unofficial channels have been opened up to help propagate the work of side events. Since SBSTA/SBI 12 in 2000, with funding from the secretariat, the Earth Negotiations Bulletin has reported on selected side events, providing a summary of discussions, photographs, contact details, and links to relevant websites. The secretariat also provides an internet webcast on selected side events. At COP 9, the secretariat introduced an additional facility for the organizers of side events to place their presentations on the secretariat website, thereby making it available to a global audience.

Demand to hold side events now outstrips the availability of time slots in the official schedule. At recent COP sessions, many NGOs have therefore decided to organize their own meetings, in their own venues, outside the official schedule, thereby also abandoning the strictures of that schedule (e.g. time slot of only two hours). ENGOs, for example, held an 'adaptation day' at both COP 8 and COP 9, while BINGOs held several events in their own meeting rooms at COP 9. In addition to side events focused mostly on discussion and debate, NGOs – mostly ENGOs – occasionally stage more visible, attention-attracting activities. These include, for example, the 'fossil of the day' award, a piece of coal being given to the country deemed to have taken the most environmentally regressive stance that day. Book launches and receptions also take place as unscheduled events, while other, more innovative examples at COP 9 included a fashion show and a mime artist.

In tandem with side events, the secretariat provides space in the conference centre for NGOs and IGOs (and also parties) to set up and run exhibit stalls. These might consist of distribution points for working papers, pamphlets or other promotional literature, exhibitions of posters or placards illustrating projects or programmes, or even include video presentations or the occasional talk. Like side events, these exhibits provide a further opportunity for NGOs to showcase their work and their views.

Unofficial vehicles

Some of the most important vehicles available to NGOs to lobby delegates and influence the climate change negotiations are largely outside the control of the organization of the negotiation process although, crucially, they are facilitated by its existence. A small number of parties (e.g. Canada) include NGO representatives on their delegations, providing a direct channel for NGOs to input into the position of the party concerned and thereby contribute to the negotiation process. Others (e.g. the EU and US) convene regular, often daily, meetings with their domestic constituencies.

Aside from these more formal links, NGOs make extensive use of the densely populated theatre provided by the two-week negotiating sessions. The presence of several thousand individuals working on climate change in a single conference centre provides critical opportunities for intensive interaction, lobbying, networking and the exchange of ideas and information. Several interviewees pointed to the importance of such informal lobbying. An NGO interviewee stated, 'you get at the people you want to get to, hopefully on a one-to-one basis, irrespective of the process ... The corridors are the most important places to find people ... where you can bump into them.' A government interviewee agreed that 'definitely the best kinds of interaction are in the corridors'. NGOs also have the opportunity to organize press briefings, making use of the often high media presence at key COP sessions. CAN-International, for example, held a press briefing on most mornings during COP 9.

Summary and concluding remarks

Overall, the climate change regime is widely viewed as relatively open and transparent to participation by NGOs, providing expansive access and multiple formal and informal opportunities to feed into the process. Of course, the right of NGOs to input into the negotiations is not on a par with that of parties; only parties can negotiate and take decisions, so there are many occasions where arenas will be closed to NGOs. On balance, however, the access granted to NGOs has broadened over the lifetime of the climate change regime. In the post-Marrakesh era, this broadening has also been accompanied by attempts to formalize the involvement of NGOs, codifying established practices.

This is certainly one area where the climate change regime has learnt over time and developed more progressive means of harnessing the (usually constructive) resources that NGOs can bring to the negotiation process. The secretariat has been at the forefront of developing and implementing informal rules and practices to manage NGO involvement. The admission process, the constituency system, the issuing of NGO written inputs as web-only documents, and the drafting of a code of conduct, have all been initiatives of the secretariat, that have been subject to only the most minimal scrutiny on the part of the subsidiary bodies. This hands-off approach by the parties has allowed the secretariat to devise and implement pragmatic ways of dealing with the great interest in the climate change regime on the part of NGOs. A degree of formalization, however, may now be necessary, given the large numbers of active NGOs, and the multiplicity of different arenas in which they wish to participate.

Conclusions: 12 Key Insights

It's a faulty process, but it's the only one we have (interview).

The objective of this book was always a simple and humble one: to improve understanding of the organization of global negotiations, and how organizational factors can impact on the course of a negotiation process. The preceding chapters discussed in detail organizational factors at play in the climate change negotiations, throwing light onto the usually unseen organizational world that lies behind the visible face of the negotiation process. It is now time to distil some more general lessons from the detailed analysis presented in those chapters. This concluding chapter takes on this task, by drawing together 12 key insights from the experience of the climate change negotiations that could provide useful lessons for other global negotiation processes.

1 The actions of the organizers are key

While formal rules of procedure and established practices provide important building blocks for the organization of a negotiation process, the actions of the organizers – the presiding officers, bureau and secretariat – are absolutely key to the effective management of the negotiations. The fact that the rules and practices of the climate change regime, and indeed other regimes, tend to be rather standard in the international arena places a premium on the ability of the organizers to interpret, implement and improvise upon those rules and practices to adapt them to the needs of the particular regime and negotiation process.

2 Unity and continuity are important

The climate change negotiations appear to have fared better under a single presiding officer, a single negotiating body and a single secretariat team for a negotiating round (the Kyoto Protocol negotiations) than under multiple presiding officers, negotiating bodies, and secretariat teams (the post-Kyoto negotiations). Where the complexity of the negotiation process requires multiple institutional actors, there should be strong and continuous communication between them, with overall responsibility vested in a senior political figure throughout the negotiating round.

3 Presiding officers should be able to supply strong process-oriented leadership

The single factor that can make the greatest difference to a negotiation process is the presence of an effective presiding officer. Although different types of negotiations and negotiating rounds require different skills, in all cases, objectivity, determination, strong decision-making capabilities, and longstanding negotiating experience in the particular regime are key factors in facilitating effective chairing and process-oriented leadership. A real problem in this regard, common throughout the international arena, is that the election of presiding officers is usually determined more by political considerations than suitability for the post in terms of skills and attributes.

4 Competent support from the secretariat is vital

Given that the appointment of effective presiding officers cannot be guaranteed, it is particularly vital that a regime enjoys the support of a competent and stable secretariat. The secretariat should be able to compensate for any chairing weaknesses on the part of the presiding officers, and also generate trust in the process by respecting the 'veil of legitimacy'. Secretariat staff should possess a balance of procedural and technical expertise that enables them to formulate strategic means of organizing the negotiation process, while also providing effective substantive, technical and logistical support. A particularly important role for the secretariat, as discussed further below, is to serve as the institutional memory of a regime, which can in turn improve its ability to organize the negotiation process effectively.

5 A balance must be struck between procedural equity/transparency and efficiency

Balancing procedural equity/transparency on the one hand, and efficiency on the other, is a central aim for the organization of the negotiation process. Success in this regard depends on the application of most of the most important rules and established practices most of the time, but knowing when the efficiency benefits of relaxing the rules would outweigh the legitimacy benefits of applying them. Placing too much emphasis on efficiency (e.g. focusing too much on closed, limited membership groups) can erode the legitimacy of the negotiation process, while excessive concern with procedural equity/transparency can jeopardize the reaching of agreement, or paradoxically provoke the key players to retreat into highly inequitable, untransparent private deals. The inability to achieve a stable, acceptable balance between these factors is widely acknowledged as a trigger to the breakdown of the COP 6 negotiations.

6 Procrastination must be resisted, and bargaining and deal-making promoted

Another central aim of the negotiation process must be to resolutely counter the inherent tendency of delegates to backload negotiations, stagnate in the initial exploratory stage, and delay the start of intensive bargaining. Although organizational tools – such as the use of texts, negotiating arenas, verbal exhortation, symbolic markers and reinforcement of the deadline – can rarely avoid the marathon, late night meetings that characterize finales of major negotiating rounds, they can ensure that the workload for those last minute negotiations is not insurmountable. Again, the failure to resist procrastination is commonly cited as a factor in the collapse of COP 6.

7 Texts must be actively managed

Texts are extremely important tools for a global negotiation, helping to disentangle the complexity arising from the multitude of issues on the negotiating table. Well-drafted texts can introduce a unified structure to diverse approaches, draw out common strands from disparate proposals, highlight key questions and options, draw attention to interlinkages, and otherwise bring some kind of order to the mass of differing proposals put forward by parties. A well-managed textual development process, whereby texts become increasingly streamlined and focused, is a critical element of an effective negotiation process, especially in terms of time management and resisting procrastination. Overlong, intricate texts that stagnate over time and are difficult to work with are usually indicators, as well as triggers, of a dysfunctional negotiation process that overburdens the negotiation finale. The presiding officer must therefore take firm charge of the textual development process, ensuring that a new text is produced for each negotiating session that registers and promotes a meaningful step forwards in the negotiation process.

8 Suitable forums are needed for exploration, as well as bargaining and deal-making

While bargaining and deal-making are the core activities of a negotiator, a negotiation process should also provide forums where delegates can engage in exploration and open discussion of issues, without the political constraints and commitments implied in bargaining. To maximize their usefulness, such complementary forums – so-called as they 'complement' the negotiating forums – should offer opportunities for actors other than the usual negotiators to also share their ideas and perspectives. These actors include NGOs (environmental, business, research and others), IGOs and scientists, and also ministers, whose political power and broader perspective allow them to make important contributions to the negotiation process. Such complementary forums can also help promote a more cooperative atmosphere, allowing delegates to speak their minds outside the political taboos and conflicts of the negotiating arenas. The climate change regime has been rather successful in opening up complementary forums, such as ministerial roundtables, workshops, and side events. In common with other regimes, however, the energy, creativity and openness often displayed in these forums needs to be better connected to the official negotiation process.

9 Procedural obstruction must be minimized

Where obstructionist delegates are present, the organizers of the negotiation process should strongly resist attempts at delaying the negotiations through procedural tactics, while respecting legitimate procedural complaints. An important element of this is strong and judicious decision-taking by the presiding officer.

10 Different negotiating contexts have different organizational needs and opportunities

Different negotiating rounds take place within different geopolitical and historical contexts. Each negotiating round thus has its own organizational needs and opportunities, while the organization of previous negotiating rounds can have repercussions on the present. The post-Kyoto negotiations, for example, had a greater need for exploration in complementary forums (given the innovative nature of their subject matter), while also facing a more conflictual context (e.g. US repudiation of the Kyoto Protocol) and less tolerance of procedural breaches (such as multiple parallel meetings) from many parties. In turn, lingering unhappiness with the closed ministerial roundtable at COP 2 and the friends group at COP 4 had long-lasting ramifications on the acceptable options open to presiding officers post-Kyoto. Because each negotiation will face its own specific challenges, it is not possible to apply a single, comprehensive organizational formula to all negotiations. This does not mean that no organizational lessons can be drawn from the experiences of individual negotiations – indeed these conclusions suggest a number of broadly applicable insights – only that a specific organizational strategy that works in one negotiation may need to be adapted so as to be effective in a different context.

11 Innovate with caution

Familiarity and stability are important contributors to efficiency, and organizational boundaries should not be pushed too far too fast. While long term learning is very important – as we see below – excessive innovation from meeting to meeting is likely to backfire, undermining the stable expectations of delegates as to how the negotiations are conducted and fomenting mistrust. Organizational innovations, however brilliant in theory, will flounder if parties are unable to make sense of them. The risk of this happening is exacerbated by the heterogeneity of parties, each with their own ways of thinking, along with the presence of obstructionists, only too happy to exploit possible misunderstandings. The differing approaches to negotiating texts and arenas at COP 6, for example, arguably went too far into the unknown; when tried again at COP 6 (part II), their greater familiarity meant that they enjoyed greater success.

12 A continuous negotiation process requires gradual learning and adaptation

Notwithstanding the dangers of excessive innovation, continuous gradual learning is very important for the sustained success of a continuous negotiation process. Learning is not an abstract concept that somehow permeates a regime, but, like the organization of a negotiation process itself, must be consciously undertaken by concrete actors. The prime actor in this respect is the secretariat, which serves as the institutional memory of the regime. The secretariat should be able to monitor and assess experiences in the organization of negotiations, applying these experiences to new negotiation rounds, and communicating the lessons learned to new staff so that they are not forgotten. The climate change secretariat has been rather active in reflecting on the negotiation process: the Executive Secretary, for example, prepared recommendations on ways of improving the negotiation process in the wake of complaints over lack of transparency and late night negotiations at COP 4 (see FCCC/SBI/1999/2). Parties themselves, however, tend to be resistant to change, with mistrust a particularly important factor. Developing countries are often particularly reluctant to consider alternatives outside the status quo, for fear of losing out on procedural equity and transparency safeguards. Moreover, many organizational factors (not to mention political dynamics) are endemic to the UN system, which is particularly resistant to change.

Overall, the climate change regime has clearly shown itself capable of learning and responding to changing circumstances, albeit ponderously and within the confines of the wider UN and political context. The proliferation of inter-sessional workshops, for example, addressed concerns over a crowded agenda, the ministerial roundtables responded to disenchantment with the general debate, the evolution of practices on NGOs answered demands for more structured involvement of these actors, and the greater use of shuttle diplomacy emerged from the dysfunctions of friends groups. Almost every aspect of the organization of the climate change negotiations has seen evolution since COP 1. Interestingly, at the time of writing, the SBI had initiated a formal review of the organization of the climate change process, with a view to ensuring that the Convention bodies could 'work as efficiently and effectively as possible' (see SBI 20 report, paragraph 94). Most of the innovations to date, however, have emerged spontaneously or on the initiative of the secretariat, rather than through negotiated decisions by the parties. Indeed, organizational reviews by the parties have, so far, tended to lead to rather conservative outputs or the formalization of established practices and secretariat initiatives. This brings us back to the first insight of this chapter – the important role played by the organizers of the negotiation process.

*****

The main message of this book is that negotiations can be organized more or less well, and this can influence a negotiation for good or ill. The organization of the negotiation process may be only one ingredient among the many factors

– power, interests, science, geopolitics, public opinion, individual personalities, history, serendipity – that together make up a negotiation and lead it to success, or not. It is, however, one of the few factors that is directly open to collective manipulation. Although the best organized process in the world cannot resuscitate a doomed negotiation if political will is lacking, the way in which a negotiation process is organized can make the difference between success and failure in a complex negotiation where the overwhelming majority of parties genuinely do want to reach a substantively meaningful agreement. This study has taken a first step in promoting a better understanding of the organization of global negotiations, in order to ensure that opportunities to advance progressive global governance are reaped to the maximum extent and are not thwarted on mere organizational grounds.

Notes

Chapter 2

1 On different negotiating techniques, see Iklé (1964), Sjöstedt et al (1994) and Zartman (1994).
2 On the importance of culture in negotiation, see Faure (1999) and Zartman (1999).
3 'The more items at stake can be divided into goods valued more by one party (or parties) than they cost to the other(s) and goods valued more by the other party (or parties) than they cost to the first, the greater the chances of a successful outcome'; cited in Sjöstedt et al, 1994, p8).
4 In the interests of simplicity, the remainder of this chapter will refer only to a 'Conference of the Parties' (COP).

Chapter 3

1 Melinda Kimble, senior US negotiator on the climate change negotiations, conversation with the author during the first Extraordinary Conference of the Parties to the CBD, Cartagena, 1999.
2 On the science of climate change and its impacts, see IPCC (2001a and b).
3 A fuller discussion of conventional economic analyses in the context of climate change can be found in IPCC (1996:ch.8, 9), IPCC (2001c:ch.3, 8, 9, Technical summary), IPCC (2001d, Question 7), and Grubb et al (1999:Appendix II).
4 This pervasiveness can be compared to the problem of ozone depletion, where the production of chlorofluorocarbons (CFCs) was dominated by only a few companies in just over 20 countries when the Montreal Protocol was adopted, with one company alone accounting for about a quarter of world production (Benedick, 1991; Oberthür, 1999).
5 For example, 120 years for N_2O, thousands of years for some long lived gases, such as sulphur hexafluoride (both of which are covered by the Kyoto Protocol).
6 For a fuller discussion of intergenerational equity in the context of global environment problems, see Weiss (1989).
7 For a history of the climate change issue, and how it rose up the international political agenda, see Bodansky (1993) and Paterson (1996).
8 For a detailed explanation and analysis of the UNFCCC, see Yamin and Depledge (2004).
9 The COP has also established three specialized bodies. The Consultative Group of Experts on National Communications from Non-Annex I parties; the Expert Group on Technology Transfer; and the Least Developed Countries Expert Group. These bodies are not discussed in this book, given that they are involved in implementation, analytical and capacity-building activities, rather than actual political negotiation. The three specialized bodies of the Kyoto Protocol – the Article 6 Supervisory Committee, the CDM Executive Board, and the Compliance Committee – are similarly not discussed in this book for the same reasons.
10 For a detailed explanation and analysis of the Kyoto Protocol, see Yamin and Depledge (2004).
11 Although the precise mandates for the work on JI, the CDM and emissions trading

varied considerably, they were all eventually interpreted as basically requiring wide-ranging work on the operationalization of the mechanisms.

12 This refers to a proposal by Iceland to consider the situation of small countries whose emissions baselines are so low, that a single 'project', e.g. the building of an aluminium smelter, could cause them to overshoot their target, even if based on the most environmentally sound technology. See Yamin and Depledge (2004).

13 See, for example, the Montreal Protocol on Substances that Deplete the Ozone Layer (Montreal Protocol, 1987) and the Convention on Biological Diversity (CBD, 1992).

Chapter 4

1 According to rule 27.1, the rules of procedure 'apply mutatis mutandis to the proceedings of the subsidiary bodies'.

2 For example, the flexibility mechanisms, LULUCF, and Articles 5, 7 and 8 (methodological issues, reporting and review).

3 But not for the more private process of informal consultations, where a single Chair is usually appointed, see Chapter 9.

4 This issue in the climate change regime refers to the adverse effects of both climate change, and actions taken to mitigate climate change.

Chapter 5

1 The joint working group on compliance (1999–2000) did not have a bureau.

2 During the negotiations on the Convention under the Intergovernmental Negotiating Committee (INC), the Bureau consisted of five members (President, Rapporteur and three Vice-Presidents), one for each official UN regional group.

3 This rule was waived at COP 6, when a Bureau member was elected for a third term, to permit him to serve for a (permitted) second term as SBI Chair (COP 6 report part I, paragraph 43). The rotation requirement, however, means that COP Presidents have never been re-elected (Yamin and Depledge, 2004).

4 'Credentials' are written attestations of members of a delegation, and are issued by the Head of State, Head of Government, or Minister of Foreign Affairs.

Chapter 6

1 These programmatic activities include coordinating the review of national communications, support to LDCs, and public outreach initiatives.

2 This includes staff financed by the Trust Funds and the Bonn Fund, as well as the core budget. See FCCC/SBI/2003/12. Although not all posts were filled, most of the vacant posts were temporarily occupied by short term staff (see also Yamin and Depledge, 2004).

3 The situation of the ozone secretariat differs, of course, from that of the climate change secretariat, in that the former is located within UNEP, which has a public information mandate and structure to deliver. Nonetheless, the contrast is a striking and pertinent one.

4 Climate change posters were produced for the 1992 Earth Summit, but since then, only occasionally, by COP host governments rather than the secretariat. A calendar was produced by the secretariat in 2002.

Chapter 7

1 For readability, unless otherwise specified, references to 'rules' in the remainder of this chapter will also encompass informal practices.
2 Some scholars have, however, raised questions as to the equitable nature of such sovereign equality, noting, for example, that it takes no account of population size, contribution to the problem, or nature of commitments under the regime (e.g. see Franck, 1995).
3 This adoption was not straightforward, however, partly due to excessive zeal on the part of the secretariat to correct problems in the text (see Chapter 11) and partly due to substantive and procedural concerns raised by the Russian Federation (see also Chapter 9).
4 Iran, the G-77 Chair for COP 6 (part II), was also an OPEC member, yet presided over a more compliant Group. At that time, however, the desire of most developing countries, Iran especially, to uphold multilateralism in the face of US repudiation of the Kyoto Protocol undoubtedly overrode tendencies towards procedural opportunism. Iran is also known for its more conciliatory stance among OPEC members.

Chapter 8

1 '...so long as legitimate US interests [are] protected'. See FCCC/CP/2001/MISC.4. However, the US has, so far, taken a relatively lax view in this respect. It did not, for example prevent agreement on the Compliance Committee, despite having opposed its composition and decision-making procedures. These procedures are based on a majority of non-Annex I parties, and with a majority of non-Annex I party members (as well as Annex I party members) required to pass a vote, which the US saw as a highly unwelcome precedent for future international institutions.
2 There are many examples of other multilateral negotiations where the definition of consensus has been contested. Indeed, as Werksman (1999, p7) notes, 'institutions do ... either through rules or practice, develop their own highly contextual definitions of consensus'.
3 A similar demand from Iceland was considered more sympathetically, given Iceland's low emissions baseline, small emissions per capita and clean energy economy.
4 On the 'Iceland issue', see Chapter 3.
5 In essence, the proposal would have granted Canada credit on account of its sale of cleaner energy to the US. The proposal, however, evolved considerably throughout its consideration in the SBSTA. For more detail, see the submission by Canada in FCCC/SBSTA/2003/MISC.7.
6 The resolution of other issues, in the meantime, also helped to alleviate US concerns.
7 Other delegations raised concerns about certain points, but stopped short of lodging a formal objection.
8 The Geneva Ministerial Declaration is contained in the annex to part II of the COP 2 report, the texts of statements made in connection with the Declaration are contained in Annex IV to part I of the report.

Chapter 9

1 In English, unless the meeting is being held in a Francophone country.
2 By watering down Canada's proposal to simply taking note of that country's intention to convene an informal meeting on the topic. For a brief discussion of the issue, see Chapter 8.

3   See the example relating to LDCs above.

4   These examples are discussed further in Chapter 8 and, in the case of general commitments, also in Chapter 4.

5   At COP 6 (part II) the 'last days' were in fact at the close of the first week, as the intention was to reach a political agreement in the first week, and engage in (less intense and controversial) technical negotiations in the second week.

6   At COP 6 (part II) the final agreement went straight to plenary, partly to ensure there was less time for it to unravel, and partly because the friends group had proved relatively ineffective in its work.

7   Cluster 1 on capacity-building, technology transfer, adverse effects and financial mechanisms, and guidance to the GEF; and cluster 4 on compliance, accounting reporting and review, and policies and measures.

8   It is important to be aware that the term 'informal consultations' is a very versatile one and, in addition to describing the specific informal arena discussed here, is also commonly used as a euphemism to refer to any small, private meeting, from friends groups to bilateral talks between the COP President (or any other presiding officer) and individual parties or negotiating coalitions (Yamin and Depledge, 2004). Both these latter examples are discussed in more detail below.

9   The former issue was finally resolved at COP 7, the latter emerged at COP 7 and is ongoing at the time of writing.

10 See Chapter 4 on the use of the term 'facilitator'.

11 For example, at SBSTA/SBI 10 in 1999 and at COP 9 in 2003, respectively, with differing distributions between informal consultations and contact groups.

12 The three flexibility mechanisms were also clustered together for more hard headed political reasons. Parties and coalitions placed differing importance on the individual mechanisms, and a good compromise found in the BAPA was thus to address them together, to reassure all parties that none would be 'left behind'.

13 On the effects of 'negotiation by exhaustion', see Chapter 12.

14 See Sebenius (1984), Sanders (1989) and Benedick (1991) on negotiations on the Law of the Sea, the Third Review Conference of the Non-Proliferation Treaty, and the Montreal Protocol, respectively.

15 See photo on http://www.iisd.ca/climate/cop6bis/20july.html.

16 Interestingly, the initial intent behind this request was to appoint a 'facilitator' who would help the Polish COP President in preparations for COP 6, but this was rejected especially by developing countries, who feared it could lead to a closed negotiation process, from which they might be excluded. See statement by Saudi Arabia calling for negotiations to stay within the purview of the subsidiary bodies and guarding against 'any external interference in the existing structures'. COP 5 report, part I, paragraph 48.

17 The issue of differentiation referred to whether Annex I parties should all have to meet the same emission target, or whether different individual targets should be allocated, and if so, how.

18 Calculation by the author based on list of speakers held with the secretariat.

Chapter 10

1   In addition to the AGBM roundtables held under the Kyoto Protocol negotiations, the other subsidiary bodies also convened occasional complementary forums as part of their own agenda of work.

2   For the full title, date and venue of each workshop referenced in this chapter, see the list of workshops on the secretariat website: http://unfccc.int/sessions/workshops.html.

3   Reports may be found at www.iisd.ca/linkages.

Chapter 11

1 The large volume of submissions also included inevitably lengthy national data tables on the sector. While not proposals as such, meaningful and informed participation in the negotiation process still required negotiators to study these tables.
2 Whether the use of the flexibility mechanisms should be supplemental to domestic action, and if so, how much, and how this should be enforced.
3 The three flexibility mechanisms – joint implementation, the CDM and emissions trading – plus the issues of accounting and registries.

Chapter 12

1 Consideration of the specific issue of sinks began late in the Kyoto Protocol negotiation process because neither the secretariat nor Chair Estrada had appreciated the importance of this issue. It was only when drafting the Chair's text (see Chapter 11), and when a landmark proposal was received from New Zealand, that the import, and complexity, of the topic became apparent.
2 This refers to the determination on the part of the EU that it should be allowed to fulfil its emission targets jointly, eventually codified in Kyoto Protocol Article 4.
3 Borrowing refers to a US proposal whereby a party could use up some of its allowed emissions for the second commitment period already in the first period.
4 The role of the COP Presidents was much less important in the Kyoto Protocol negotiations, where Chair Estrada acted as the single leader (see Chapter 4).
5 The negotiations on the UNFCCC (Mintzer and Leonard, 1994a), CBD (McConnell, 1996), the Stockholm Convention on Persistent Organic Pollutants (ENB, 2000c) and both the first and second negotiating rounds of the Cartagena Biosafety Protocol (Bail et al, 2002 and ENB, 2000a) all experienced exhausting, late night negotiation finales.

Chapter 13

1 The term 'minister' is used to denote an individual who is part of the government in power in a state and who has responsibility for taking political decisions relating to the negotiations. A minister so defined may (e.g. UK) or may not (e.g. US) be directly elected, depending on the political system.
2 The exception was COP 6 (part II), where the high-level segment was held in the middle of the session.
3 Speeches by some hosting heads of state have been summarized, the speech made on behalf of the King of Morocco at COP 7, for example, and the speech by the Chancellor of Germany at COP 1.
4 Except COP 6 (part II), which was convened only to conclude the negotiations that broke down at COP 6.
5 The Geneva Ministerial Declaration, however, did make an important contribution to the Kyoto Protocol negotiations. For more on the Declaration, see Chapters 8 and 12.
6 For a discussion of how the consensual and benign political atmosphere of the COP 5 roundtable may actually have hampered the post-Kyoto negotiations, see Chapter 12.

Chapter 14

1 This chapter does not directly consider IGOs or UN agencies, as participation by these

organizations tends to be more passive (e.g. monitoring proceedings) or project-based (e.g. conducting adaptation activities in the field), rather than focused on making inputs to the negotiations. However, the rules and practices relating to NGOs discussed in this chapter do apply, for the most part, also to IGOs and UN agencies, with important differences highlighted. Participation by the media is governed only skeletally by the organization of the negotiation process and is therefore not discussed here.

2   A vast literature has emerged examining the role of NGOs in international regimes in general, and the climate change negotiations in particular. See, for example, Chatterjee and Finger (1994), Albin (1999), Newell (2000), Carpenter (2001), Yamin (2001).

3   See, for example, the rules of procedure for the ozone regime (UNEP, 2003) and the CBD (CBD, 1994).

4   The US statement concerned not only the participation of NGOs at workshops, but also the attendance of parties to the *Convention* (but not the protocol) as observers to expert groups, notably the CDM-Executive Board. See FCCC/SBI/2002/MISC.8.

5   One of the proposals floated as part of the pre-Kyoto debate on NGO participation was for NGOs to be required to 'declare support for the aims of the Convention, for example, its objective and principles' (FCCC/SBI, 1997d, p3). Such a proposal was, unsurprisingly, strongly denounced by NGOs of an obstructionist bent (e.g. Ecologic, 1997).

6   This was formally acknowledged by the SBI in the conclusions of its 20th session in 2004. See ISBI 20 report, paragraph 102.

7   Although some European ENGOs (e.g. Climate Network Europe) sometimes included EIT nationals on their delegations.

8   Agenda 21 (1992) was a key output of the 1992 Earth Summit, setting out a 40-chapter 'blueprint' for advancing towards sustainable development.

9   At this point, the US was in essence relaying the complaints of certain (US-based) BINGOs, who were not satisfied with the use of the International Chamber of Commerce as their constituency focal point. The focus of US concerns later switched in 2004 to the newly established RINGO constituency and the basis for its recognition.

10 Available at http://unfccc.int/resource/ngo/coc_guide.pdf.

11 The rules of procedure state that 'Meetings of the Conference of the Parties shall be held in public, unless the Conference of the Parties decides otherwise'. The rules of procedure state that subsidiary body meetings shall be held in private unless the COP decides otherwise, but in accordance with established practice in climate change regime, this means that admitted observers may attend.

12 Author's calculation, based on cassette recording on file with secretariat. Many speakers did not identify themselves, and could have been from NGOs.

13 Although 'side events' were termed 'special events' until SBSTA/SBI 18 in 2003.

References

Primary material: Official documentation, statements and related material

Decisions cited may be found in part II of the report of the relevant COP session. Decision 18/CP.4, for example, can be found in the COP 4 report, part II. UN resolutions cited are available at www.un.org.

AGBM 2 (1995) Second session. *Exchange between Estrada and delegates*. AGBM 2:0059, cassette recording held with secretariat

AGBM 4 report, *Report of the Ad Hoc Group on the Berlin Mandate at its fourth session, Geneva, 11–16 July 1996*. Document FCCC/AGBM/1996/8

AGBM 6 (1997) Sixth session. *Closing statement by Estrada*. Notes on file with secretariat and author

AGBM 6 report, *Report of the Ad Hoc Group on the Berlin Mandate on the work of its sixth session, Bonn, 3–7 March 1997*. Document FCCC/AGBM/1997/3

AGBM 7 (1997a) Seventh session. *Intervention by China during the non-group on QELROs*. AGBM 7:0320, cassette recording held with secretariat

AGBM 7 (1997b) Seventh session. *Intervention by Iran*. AGBM 7:0317, cassette recording held with secretariat

AGBM 7 (1997c) Seventh session. *Intervention by Estrada*. AGBM 7:0363, cassette recording held with secretariat

AGBM 8 (1997a) Eighth session. *Opening statement by the US*. AGBM 8: 0367, cassette recording held with secretariat

AGBM 8 (1997b) Eighth session. *Opening statement by Estrada*. AGBM 8:0385, cassette recording held with secretariat

AGBM 8 (1997c) Eighth session. *Opening statement by Zimbabwe*. AGBM 8:0385, cassette recording held with secretariat

AGBM 8 (1997d) Eighth session. *Exchange between Estrada and delegates*. AGBM 8:0459, cassette recording held with secretariat

AGBM 8 (1997e) Eighth session. *Closing statement by Estrada*. AGBM 8:0477, cassette recording held with secretariat

AGBM 8 (1997f) Eighth session, part II. *Intervention by the Russian Federation*. Notes on file with secretariat and author

AGBM Bureau (1997) *Speaking notes for Estrada for use at the AGBM Bureau, AGBM 8*. On file with secretariat and author

COP 1 report, part I, *Report of the Conference of the Parties at its first session held at Berlin from 28 March to 7 April 1995. Part I: Proceedings*. Document FCCC/CP/1995/7

COP 1 report, part II, *Report of the Conference of the Parties at its first session held at Berlin from 28 March to 7 April 1995. Part II: Action taken by the Conference of the Parties*. Document FCCC/CP/1995/7/Add.1

COP 2 report, part I, *Report of the Conference of the Parties at its second session held at Geneva from 8 to 19 July 1996. Part I: Proceedings*. Document FCCC/CP/1996/15

COP 2 report, part II, *Report of the Conference of the Parties at its second session held at Geneva from 8 to 19 July 1996. Part II: Action taken by the Conference of the Parties*. Document FCCC/CP/1996/15/Add.1

COP 3 report, part I, *Report of the Conference of the Parties on its third session, held at Kyoto from 1 to 11 December 1997. Part I: Proceedings*. Document FCCC/CP/1997/7

COP 3 report, part II, *Report of the Conference of the Parties on its third session, held at Kyoto from 1 to 11 December 1997. Part II: Decisions adopted by the Conference of the Parties*. Document FCCC/CP/1997/7/Add.1

COP 4 report, part I, *Report of the Conference of the Parties at its fourth session held at Buenos Aires from 2 to 14 November 1998. Part I: Proceedings*. Document FCCC/CP/1998/16

COP 4 report, part II, *Report of the Conference of the Parties at its fourth session held at Buenos Aires from 2 to 14 November 1998. Part II: Action taken by the COP at its fourth session*. Document FCCC/CP/1998/16/Add.1

COP 5 report, part I, *Report of the Conference of the Parties on its fifth session, held at Bonn from 25 October to 5 November 1999. Part I: Proceedings*. Document FCCC/CP/1999/6

COP 5 report, part II, *Report of the Conference of the Parties on its fifth session, held at Bonn from 25 October to 5 November 1999. Part II: Action taken by the Conference of the Parties at its fifth session*. Document FCCC/CP/1999/6/Add.1

COP 6 (part I) report, part I, *Report of the Conference of the Parties on the first part of its sixth session, held at The Hague from 13–25 November 2000. Part I: Proceedings*. Document FCCC/CP/2000/5/Add.1

COP 6 (part I) report, part II, *Report of the Conference of the Parties on the first part of its sixth session, held at The Hague from 13–25 November 2000. Part II: Action taken by the Conference of the Parties at the first part of its sixth session*. Document FCCC/CP/2000/5/Add.2

COP 6 (part II) report, *Report of the Conference of the Parties on the second part of its sixth session, held at Bonn from 16 to 27 July 2001*. Document FCCC/CP/2001/5

COP 7 report, part I, *Report of the Conference of the Parties on its seventh session, held at Marrakesh from 29 October to 10 November 2001. Part I: Proceedings*. Document FCCC/CP/2001/13

COP 7 report, part II, *Report of the Conference of the Parties on its seventh session, held at Marrakesh from 29 October to 10 November 2001. Part II: Action taken by the Conference of the Parties*. Documents FCCC/CP/2001/13/Add.1–4

COP 8 report, part I, *Report of the Conference of the Parties on its eighth session, held at New Delhi, from 23 October to 1 November 2002. Part I: Proceedings*. Document FCCC/CP/2002/7

COP 8 report, part II, *Report of the Conference of the Parties on its eighth session, held at New Delhi, from 23 October to 1 November 2002. Part II: Action taken by the Conference of the Parties at its eighth session*. Document FCCC/CP/2002/7/Add.1–3

COP 9 report, part I, *Report of the Conference of the Parties on its ninth session, held at Milan from 1 to 12 December 2003. Part I: Proceedings*. Document FCCC/CP/2003/6

COP 9 report, part II, COP 9 report, part I, *Report of the Conference of the Parties on its ninth session, held at Milan from 1 to 12 December 2003. Part II: Action taken by the Conference of the Parties at its ninth session*. Documents FCCC/CP/2003/6/Add.1–2

CoW (1997a) *First meeting, 1 December*. Notes on file with secretariat and author

CoW (1997b) *Third meeting, 2 December*. Notes on file with secretariat and author

CoW (1997c) *Fourth meeting, 3 December*. Notes on file with secretariat and author

CoW (1997d) *Tenth meeting, 6 December*. Notes on file with secretariat and author

CoW (1997e) *Thirteenth meeting, 10 December*. Notes on file with secretariat and author

CoW (1997f) *Fourteenth meeting resumed, 11 December*. Notes on file with secretariat and author

Estrada, R. (1997) *Report on the work of the AGBM to the third Conference of the Parties.* On file with secretariat and author

FCCC (1992) *United Nations Framework Convention on Climate Change.* UNEP/IUC: Geneva

FCCC (1996) *Article 17.2 of the Convention: Opinion received from the United Nations Office of Legal Affairs.* 11 July 1996

FCCC (2000) *A guide to the climate change process.* Bonn: Climate Change Secretariat

FCCC press release (1996) *Climate Change negotiations enter new phase.* 13 December 1996. On file with author, and available at www.unep.ch/iuc

FCCC/AGBM/1996/10, *Synthesis of proposals by parties*

FCCC/AGBM/1997/2, *Framework compilation of proposals from parties for the elements of a protocol or another legal instrument*

FCCC/AGBM/1997/3/Add.1, *Proposals for a protocol or another legal instrument.* Report of the Ad Hoc Group on the Berlin Mandate on the work of its sixth session, Bonn, 3–7 March 1997. Addendum

FCCC/AGBM/1997/7, *Completion of a protocol or another legal instrument: Consolidated negotiating text by the Chairman*

FCCC/AGBM/1997/INF.1, *Reports by the Chairmen of the informal consultations conducted at the seventh session of the Ad Hoc Group on the Berlin Mandate*

FCCC/AGBM/1997/MISC.8, *Implementation of the Berlin Mandate. Proposals from parties*

FCCC/CP/1996/1/Add.1, *Provisional agenda and annotations including suggestions for the organization of work. Ministerial roundtable*

FCCC/CP/1996/2, *Organizational matters: Adoption of the rules of procedure*

FCCC/CP/1997/2, *Adoption of a protocol or another legal instrument: Fulfilment of the Berlin Mandate. Revised text under negotiation*

FCCC/CP/1997/CRP.2, *Non-paper by the Chairman of the Committee of the Whole*

FCCC/CP/1997/CRP.4, Untitled

FCCC/CP/1997/CRP.6, *Kyoto Protocol to the United Nations Framework Convention on Climate Change. Final draft by the Chairman of the Committee of the Whole*

FCCC/CP/2000/5/Add.3 (Vol IV and V), *Texts forwarded to the resumed sixth session by the Conference of the Parties at the first part of its sixth session*

FCCC/CP/2000/INF.3 (Vol V), *Texts forwarded by the subsidiary bodies to the Conference of the Parties at the first part of its sixth session*

FCCC/CP/2001/2 and Add.1–6, *Consolidated negotiating text prepared by the President*

FCCC/CP/2001/5/Add.2, *Draft decisions on which progress was noted by the Conference of the Parties at the second part of its sixth session and which the Conference of the Parties decided to forward to its seventh session for elaboration, completion and adoption*

FCCC/CP/2001/CRP.8, *Note by the Co-Chairmen of the negotiating groups*

FCCC/CP/2001/MISC.4, *Statements made in connection with the approval of the Bonn Agreements on the implementation of the Buenos Aires Plan of Action (decision 5/CP.6)*

FCCC/CP/2001/MISC.9, *Closure of the session. Views from a party*

FCCC/CP/2003/1/Add.1, *Round-table discussions among ministers and other heads of delegation*

FCCC/CP/2003/OD/9, *Daily programme for Wednesday, 10 December 2003* (COP 9)

FCCC/KP (1997) *The Kyoto Protocol to the United Nations Framework Convention on Climate Change.* UNEP/IUC: Geneva

FCCC/Non-paper (2000a) *Non-paper from the President-designate of COP 6*

FCCC/Non-paper (2000b) *Non-paper from the President of COP 6*

FCCC/Non-paper (2000c) *Note by the President of the COP*

FCCC/Non-paper (2001) *Core elements for the implementation of the Buenos Aires Plan of Action*

FCCC/SB/1999/2, *Work programme on methodological issues related to Articles 5, 7 and 8 of the Kyoto Protocol*

FCCC/SB/1999/INF.2 and Add.1–3, *Mechanisms pursuant to Articles 6, 12 and 17 of the Kyoto Protocol. Synthesis of proposals from parties on principles, modalities, rules and guidelines*
FCCC/SB/1999/7 and Add.1, *Elements of a compliance system and synthesis of submissions*
FCCC/SB/1999/8 and Add.1, *Mechanisms pursuant to Articles 6, 12 and 17 of the Kyoto Protocol. Synthesis of proposals from parties on principles, modalities, rules and guidelines*
FCCC/SB/2000/1, *Note by the Co-Chairmen of the Joint Working Group on Compliance*
FCCC/SB/2000/3, *Mechanisms pursuant to articles 6, 12 and 17 of the Kyoto Protocol. Text for further negotiation on principles, modalities, rules and guidelines*
FCCC/SB/2000/4, *Mechanisms pursuant to articles 6, 12 and 17 of the Kyoto Protocol. Consolidated text on principles, modalities, rules and guidelines*
FCCC/SB/2000/7, *Proposals from the Co-Chairmen of the Joint Working Group on Compliance*
FCCC/SB/2000/10, Add.1–4, *Mechanisms pursuant to articles 6, 12 and 17 of the Kyoto Protocol. Text by the Chairmen*
FCCC/SB/2000/11, *Text proposed by the Co-Chairmen of the Joint Working Group on Compliance*
FCCC/SBI/1997/11, *Arrangements for intergovernmental meetings*
FCCC/SBI/1997/14/Add.1, *Mechanisms for consultation with non-governmental organizations: The participation of NGOs in the Convention process*
FCCC/SBI/1999/2, *Arrangements for intergovernmental meetings*
FCCC/SBI/2000/10/Add.2, *Proposals from the Co-Chairmen of the JWG*
FCCC/SBI/2002/MISC.8, *Effective participation in the Convention process. Submissions from parties*
FCCC/SBI/2003/12, *Interim financial performance for the biennium 2002–2003. Income and budget performance as at 30 June 2003*
FCCC/SBI/2003/15/Add.1, *Programme budget for the biennium 2004–2005*
FCCC/SBI/2003/INF.14, *An initial assessment of steps taken by non-Annex I parties to reduce emissions and enhance removals of greenhouse gases*
FCCC/SBI/2004/5, *Promoting effective participation in the Convention process*
FCCC/SBSTA/1999/5, *Land use, land-use change and forestry. List of policy and procedural issues related to Article 3.3 and 3.4*
FCCC/SBSTA/2000/9, *Land use, land-use change and forestry. Consolidated synthesis of proposals made by parties*
FCCC/SBSTA/2000/10/Add.1, *Mechanisms pursuant to Articles 6, 12 and 17 of the Kyoto Protocol. Consolidated text on principles, modalities, rules and guidelines*
FCCC/SBSTA/2000/10/Add.2, *Land use, land-use change and forestry. Recommendation by the SBSTA*
FCCC/SBSTA/2000/12, *Land use, land-use change and forestry. Text by the Chairman*
FCCC/SBSTA/2003/MISC.7, *Issues relating to cleaner or less-greenhouse-gas-emitting energy. Submissions from parties*
FCCC/TP/2000/2 *Tracing the origins of the Kyoto Protocol: An article-by-article history*
FCCC/WEB/2002/6, *Review of the financial mechanism, Submission from non-governmental organizations*
FCCC/WEB/2002/12, *Views from organizations on issues related to modalities for the inclusions of afforestation and reforestation project activities under the clean development mechanism in the first commitment period*
FCCC/WEB/2002/13, *Effective participation in the Convention process. Submission from a non-governmental organization*
FCCC/WEB/2002/14, *Article 6 of the Convention. Submission from a non-governmental organization*
SBI 8 report, *Report of the Subsidiary Body for Implementation on its eighth session, Bonn, 2–12 June 1998.* Document FCCC/SBI/1998/6
SBI 12 report, *Report of the Subsidiary Body for Implementation on its twelfth session, Bonn,*

12–16 June 2000. Document FCCC/SBI/2000/5

SBI 18 report, *Report of the Subsidiary Body for Implementation on its eighteenth session, held at Bonn, from 4 to 13 June 2003.* Document FCCC/SBI/2003/8

SBI 20 report, *Report of the Subsidiary Body for Implementation on its twentieth session, held at Bonn, 16–25 June 2004.* Document FCCC/SBI/2004/10

SBSTA 3 report, *Report of the Subsidiary Body for Scientific and Technological Advice on the work of its third session, Geneva, 9–16 July 1996.* Document FCCC/SBSTA/1996/20

SBSTA 6 report, *Report of the Subsidiary Body for Scientific and Technological Advice on the work of its sixth session, Bonn, 28 July –. 5 August 1997.* Document FCCC/SBSTA/1997/6

SBSTA 14 report, *Report of the Subsidiary Body for Scientific and Technological Advice on its fourteenth session, Bonn, 24–27 July 2001.* Document FCCC/SBSTA/2001/2

SBSTA 16 report, *Report of the Subsidiary Body for Scientific and Technological Advice on its sixteenth session, held at Bonn, from 5–14 June 2002.* Document FCCC/SBSTA/2002/6

SBSTA 19 report, *Report of the Subsidiary Body for Scientific and Technological Advice on its nineteenth session, held at Milan, from 1–9 December 2003.* Document FCCC/SBSTA/2003/15

ST/SGB/2002/1, *Staff rules: Staff regulations of the United Nations and Staff rules, 1 January 2002,* available at www.un.org

Tanzania (1997) *Statement by the United Republic of Tanzania for the Group of 77 and China made at the opening of AGBM 8, 22 October 1997.* On file with secretariat and author

USA (1997) *Statement of the United States of America regarding a provision on military operations for collective security, October 31 1997.* On file with secretariat and author

Other publications

Agenda 21 (1992) in Johnson, S. P. *The Earth Summit: The United Nations Conference on Environment and Development (UNCED),* Book II. London, Graham and Trotman, pp125–508

Albin, C. (1999) 'Can NGOs enhance the effectiveness of international negotiation?' *International Negotiation,* vol 4, no 3, pp371–387

Andresen, S. and Skjaerseth, J. B. (1999) *Can international environmental secretariat promote effective co-operation?* Paper presented at the International Conference on Synergies and Coordination between Multilateral Environmental Agreements, United Nations University, Tokyo, 14–16 July 1999, available at www.ias.unu.edu

Bail, C., Falkner, R. and Marquard, H. (eds) (2002) *The Cartagena Protocol on Biosafety: Reconciling Trade in Biotechnology with Environment and Development?* London, RIIA/Earthscan

Barrett, B. F. D. and Chambers, W. B. (1998) *Primer on Scientific Knowledge and Politics in the Evolving Global Climate Change Regime: COP 3 and the Kyoto Protocol.* Tokyo, United Nations University

Benedick, R. E. (1991) *Ozone Diplomacy: New Directions in Safeguarding the Planet.* Enlarged edition. Cambridge, Harvard University Press

Benedick, R. E. (1993) 'Perspectives of a negotiation practitioner', in Sjöstedt, G. (ed) *International Environmental Negotiation.* London, Sage, pp219–243

Bodansky, D. (1993) 'The United Nations Framework Convention on Climate Change: A commentary'. *The Yale Journal of International Law,* vol 18, no 2, pp451–558

Boyer, B. (1999) 'Introduction'. *International Negotiation,* vol 4, no 2, pp101–106

Brack, D. and Hyvarinen, J. (eds) (2002) *Global environmental institutions: Perspectives on reform.* Royal Institute of International Affairs, Sustainable Development Programme, available at www.riia.org

Brenton, T. (1994) *The greening of Machiavelli: The evolution of international environmental politics.* London, Earthscan

CAN (1997) 'Statement of Climate Action Network to the Third Conference of the Parties' in Taalab, A. *Rising Voices Against Global Warming.* Frankfurt, IZE, pp14–15

Carpenter, C. (2001) 'Businesses, green groups and the media: The role of non-governmental organizations in the climate change debate'. *International Affairs*, vol 77, no 2, pp313–328

CBD (1992) *Convention on Biological Diversity*, available at www.biodiv.org

CBD (1994) *Rules of procedure for meetings of the Conference of the Parties to the Convention on Biological Diversity.* Report on the first meeting of the Conference of the Parties to the CBD. Document UNEP/CBD/COP/1/17

CBD (1999) *List of participants.* Open-ended ad hoc working group on biosafety. Sixth meeting. Cartagena, 14–19 February 1999. Document UNEP/CBD/BSWG/6/INF.10

CBD (2002) *Report on the sixth meeting of the Conference of the Parties to the Convention on Biological Diversity.* Document UNEP/CBD/COP/6/20

Chatterjee, P. and Finger, M. (1994) *The Earth Brokers: Power, Politics and World Development.* London, Routledge

Dessai, S., Schipper, E. L. F., Corbera, E., Haxeltine, A., Kjellen, B., and Gutierrez, M. (2004) *Challenges and outcomes at COP 9.* Tyndall Briefing Note No 11, available at www.tyndall.ac.uk

ECO (1995) *Léman.* 22 August 1995, available at www.climatenetwork.org/eco

ECO (1996) *Science back in front.* 18 July 1996

ECO (1997a) *The Access Question.* 4 March 1997

ECO (1997b) *Conference report.* 9 December 1997

Ecologic (1997) *Who should participate?* 1 August 1997 (on file with author)

ENB (1996) *Summary Report on the Second Conference of the Parties, 8–19 July 1996*, vol 12, no 38, available at www.iisd.ca/linkages

ENB (1998) *Highlights from the fourth UNFCCC Conference of the Parties, 10 November 1998*, vol 12, no 94, available at www.iisd.ca/linkages

ENB (1999) *Summary of the fifth Conference of the Parties to the Framework Convention on Climate Change. 25 October to 5 November 1999*, vol 12, no 123, available at www.iisd.ca/linkages

ENB (2000a) *Report of the resumed session of the Extraordinary Meeting of the Conference of the Parties for the adoption of the Protocol on Biosafety to the Convention on Biological Diversity, 24–28 January 2000*, vol 9, no 137, available at www.iisd.ca/linkages

ENB (2000b) *Summary of the Sixth Conference of the Parties to the Framework Convention on Climate Change, 13–25 November 2000*, vol 12 no 163, available at www.iisd.ca/linkages

ENB (2000c) *Summary of the Fifth Session of the Intergovernmental Negotiating Committee for an International Legally Binding Instrument For Implementing International Action on Certain Persistent Organic Pollutants, 4–9 December 2000*, , vol 15 no 54, available at www.iisd.ca/linkages

ENB (2000d) *Summary of the thirteenth sessions of the subsidiary bodies of the UN Framework Convention on Climate Change, 4–15 September*, vol 12, no 151, available at www.iisd.ca/linkages

ENB (2001) *Summary of the resumed sixth session of the Conference of the Parties to the UN Framework Convention on Climate Change, 16–27 July 2001*, vol 12 no 176, available at www.iisd.ca/linkages

ENB (2002) *UNFCCC SB-16 Highlights, 6 June 2002*, vol 12, no 194

ENB (2003a) *A summary of the ninth Conference of the Parties to the United Nations Framework Convention on Climate Change, 1–12 December 2003*, vol 12, no 231

ENB (2003b) *A brief analysis of the COP-9 side-events*, vol 13, no 2

ENB (2004) *Summary of the twentieth sessions of the subsidiary bodies of the United Nations*

Framework Convention on Climate Change, 16–25 June 2004, vol 12, no 242, available at www.iisd.ca/linkages

Evensen, J. (1989) 'Three procedural cornerstones of the Law of the Sea Conference: The consensus principle, the package deal and the gentleman's agreement', in Kaufmann, J. (ed) *Effective Negotiation: Case Studies in Conference Diplomacy*. Dordrecht, Martinus Nijhoff Publishers, pp75–92

Faure, G-O (1999) 'Cultural aspects of international negotiation', in Berton, P., Kimura, H. and Zartman, I. W. (eds) *International Negotiation: Actors, Structure/Process, Values*.London, Macmillan, pp11–32

Fisher, R., Ury, W. and Patton, B. (1992) *Getting to Yes: Negotiating an Agreement Without Giving In*. Second edition. London, Random House Business Books

Franck, T. M. (1995) *Fairness in International Law and Institutions*. Oxford, Clarendon Press

Freymond, J. F. (1991) 'Historical approach', in Kremenyuk, V. A. (ed) *International Negotiation: Analysis, Approaches, Issues*. San Francisco, Jossey-Bass, pp121–134

Globe (Global Legislators Organization for a Balanced Environment) (1998) The oxen and the butterflies. Statement to COP 3, 7 December 1997. In Taalab, A. *Rising Voices Against Global Warming*. Frankfurt, IZE, pp178–180

Grubb, M. (1995) 'Seeking fair weather: ethics and the international debate on climate change'. *International Affairs*, vol 71, no 3, pp463–496

Grubb, M., Vrolijk, C. and Brack, D. (1999) *The Kyoto Protocol: A Guide and Assessment*. London, Earthscan

Grubb, M. and Yamin, F. (2001) 'Climate collapse at The Hague: What happened, why, and where do we go from here?' *International Affairs*, vol 77, no 2, pp261–276

Gupta, J. and Grubb, M. (eds) (2000) *Climate change and European leadership: A sustainable role for Europe?* Dordrecht, Kluwer Academic Press

Hampson, F. O. and Hart, M. (1995) *Multilateral Negotiations: Lessons from Arms Control, Trade and the Environment*. Baltimore, Johns Hopkins University Press

Househam, I., Hauff, J., Missfeldt, F. and Grubb, M. (1998) *Climate Change and the Energy Sector: A Country-by-Country Analysis of National Programmes. Volume 3: The Economies in Transition*. FT Management Report. London, FT Energy

Hyder, T. O. (1994) 'Looking back to see forward', in Mintzer, I. and Leonard, J. A. (eds) *Negotiating Climate Change: The Inside Story of the Rio Convention*. Cambridge, Cambridge University Press, pp201–228

Iklé, F. C. (1964) *How Nations Negotiate*. New York, Praeger

Illich, J. (1999) *The Complete Idiot's Guide to Winning Through Negotiation*. Second edition. Alpha Books. New York, Macmillan

IPCC (1996) *Climate Change 1995: Economic and Social Dimensions of Climate Change*. Contribution of Working Group III to the Second Assessment Report of the Intergovernmental Panel on Climate Change. Bruce, J. P., Lee, H. and Haites, E. F. (eds). Cambridge, Cambridge University Press

IPCC (2000) *Special Report on Land Use, Land-use Change and Forestry*. Watson, R. T., Noble, I. R., Bolin, B., Ravindranath, N. H., Verardo, D. J. and Dokken, D. J. (eds). Cambridge, Cambridge University Press

IPCC (2001a) *Climate change 2001. The scientific basis*. Contribution of Working Group I to the Third Assessment Report of the Intergovernmental Panel on Climate Change. Houghton, J. T., Ding, Y., Griggs, D. J., Noguer, M., Van der Linden, P. J., Dai, S., Maskell, K. and Johnson, C. A. (eds). Cambridge, Cambridge University Press

IPCC (2001b) *Climate change 2001. Impacts, adaptation and vulnerability*. Contribution of Working Group II to the Third Assessment Report of the Intergovernmental Panel on Climate Change. McCarthy, J. J., Canziani, O. F., Leary, N. A., Dokken, D. J. and White, K. S. (eds). Cambridge, Cambridge University Press

IPCC (2001c) *Climate change 2001. Mitigation*. Contribution of Working Group III to the Third Assessment Report of the Intergovernmental Panel on Climate Change. Metz, B.,

Davidson, O., Swart, R. and Pan, J. (eds). Cambridge, Cambridge University Press

IPCC (2001d) *Climate change 2001. Synthesis Report.* Watson, R. et al (eds). Cambridge, Cambridge University Press

Jacoby, H. and Reiner, D. (2001) 'Getting climate policy on track after The Hague'. *International Affairs*, vol 77, no 2, pp297–312

Kaufmann, J. (ed) (1989) *Effective Negotiation: Case Studies in Conference Diplomacy.* Dordrecht, Martinus Nijhoff Publishers

Kaufmann, J. (ed) (1996) *Conference Diplomacy: An Introductory Analysis.* Third edition. Basingstoke, Macmillan

Keohane, R. O. (1989) *International Institutions and State Power: Essays in International Relations Theory.* Boulder, Westview Press

Keohane, R. O. (1993) 'The analysis of international regimes: Towards a European-American research programme', in Rittberger, V. (ed) *Regime Theory and International Relations,.* Oxford, Clarendon Press, pp23–48

Keohane, R. O. and Nye, J. S. (1977) *Power and Interdependence.* Second edition. Boston, Little, Brown

Kjellen, B. (1994) 'A personal assessment', in Mintzer, I. M. and Leonard, J. A. (eds) (1994) *Negotiating Climate Change: The Inside Story of the Rio Convention.* Cambridge, Cambridge University Press, pp149–174

Kremenyuk, V. A. and Lang, W. (1993) 'The political, diplomatic and legal background', in Sjöstedt, G. (ed) *International Environmental Negotiation*, London, Sage, pp3–16

Lang, W. (1989a) 'The role of presiding officers in multilateral negotiations', in Mautner-Markhof, F. (ed) *Processes of International Negotiations.* Boulder, Westview Press, pp23–42

Lang, W. (1989b) 'The Second Review Conference of the 1972 Biological Weapons Convention', in Kaufmann, J. (ed) *Effective Negotiation: Case Studies in Conference Diplomacy.* Dordrecht, Martinus Nijhoff Publishers, pp191–204

Lang, W. (1991) 'Negotiations on the environment', in Kremenyuk, V. A. (ed) *International Negotiation: Analysis, Approaches, Issues*, San Francisco, Jossey-Bass, pp343–356

Lang, W. (1994) 'Lessons drawn from practice: Open covenants, openly arrived at', in Zartman, I. W. (ed) *International Multilateral Negotiation: Approaches to the Management of Complexity.* San Francisco, Jossey-Bass, pp201–212

Lefeber, R. (2001) From The Hague to Bonn to Marrakesh and Beyond: A Negotiating History of the Compliance Regime under the Kyoto Protocol. Hague Yearbook of International Law, vol 14, pp25–54

Litfin, K. (1994) *Ozone Discourses: Science and Politics in Global Environmental Cooperation.* New York, Columbia University Press

Martinez, J. and Susskind, L. E. (2000) 'Parallel informal negotiation: An alternative to second track diplomacy'. *International Negotiation*, vol 5, no 3, pp569–586

McConnell, F. (1996) *The Biodiversity Convention: A Negotiating History.* London, Kluwer Law International

Midgaard, K. and Underdal, A. (1977) 'Multiparty conferences', in Druckman, D. (ed) *Negotiations: Social-psychological Perspectives.* Beverly Hills, Sage, pp329–346

Miles, E. L., Underdal, A., Andresen, S. Wettestad, J., Skjaerseth, J. B. and Carlin, E. M. (2002) *Environmental Regime Effectiveness: Confronting Theory with Evidence.* Cambridge, MIT Press

Mintzer, I. M. and Leonard, J. A. (eds) (1994a) *Negotiating Climate Change: The Inside Story of the Rio Convention.* Cambridge, Cambridge University Press

Montreal Protocol (1987) Montreal Protocol on Substances That Deplete the Ozone Layer, available at www.unep.org/ozone

Mwandosya, M. (2000) *Survival Emissions: A Perspective from the South on Global Climate Change Negotiations.* Dar es Salaam, Dar es Salaam University Press and Centre for Energy, Environment, Science and Technology

Newell, P. (2000) *Climate for Change: Non-State Actors and the Global Politics of the Greenhouse.* Cambridge, Cambridge University Press

Nicolson, H. (1939) *Diplomacy*. London, Oxford University Press

Oberthür, S. (1996) 'The Second Conference of the Parties'. *Environmental Policy and Law*, vol 26, no 5, pp195–201

Oberthür, S. (1999) *Production and Consumption of Ozone Depleting Substances 1986–1997: The Data Reporting System under the Montreal Protocol*. Eschborn, Deutsche Gesellschaft für Technische Zusammenarbeit

Oberthür, S. and Ott, H. (1995) 'The First Conference of the Parties'. *Environmental Policy and Law*, vol 25, no 4/5, pp144–156

Oberthür, S. and Ott, H. (1999) *The Kyoto Protocol: International Climate Policy for the 21st Century*. Berlin, Springer Verlag

Ott, H. (2001) 'Climate change: An important foreign policy issue'. *International Affairs*, vol 77, no 2, pp277–296

Paterson, M. (1996) *Global Warming and Global Politics*. London, Routledge

Pruitt, D. G. (1981) *Negotiation Behaviour*. London, Academic Press

Putnam, R. D. (1988) 'Diplomacy and domestic politics: The logic of two-level games'. *International Organization*, vol 42, no 3, pp427–460

Raiffa, H. (1982) *The Art and Science of Negotiation*. Cambridge, Harvard University Press

Raiffa, H (1991) 'Contributions of applied systems analysis to international negotiations', in Kremenyuk, V. A. (ed) *International Negotiations: Analysis, Approaches, Issues*. San Francisco, Jossey-Bass, pp5–21

Renninger, J. P. (1989) 'The failure to launch global negotiations at the eleventh Special Session of the UN General Assembly', in Kaufmann, J. (ed) *Effective Negotiation: Case Studies in Conference Diplomacy*. Dordrecht, Martinus Nijhoff Publishers, pp231–254

Rittberger, V. (1983) 'Global conference diplomacy and international policy-making: The case of UN-sponsored world conferences'. *European Journal of Political Research*, vol 11, no 2, pp167–182

Rosenau, J. and Czempiel, E. O. (1992) *Governance Without Government: Order and Change in World Politics*. Cambridge, Cambridge University Press

Rowlands, I. H. (1995). *Politics of Global Atmospheric Change*. Manchester, Manchester University Press

Rubin, J. Z. (1991) 'The actors in negotiation', in Kremenyuk, V. A. (ed) *International Negotiation: Analysis, Approaches, Issues*. San Francisco, Jossey-Bass, pp90–99

Sanders, B. (1989) 'The Third Review Conference of the Non-Proliferation Treaty', in Kaufmann, J. (ed) *Effective Negotiation: Case Studies in Conference Diplomacy*.Dordrecht, Martinus Nijhoff Publishers, pp255–266.

Sandford, R. (1994) 'International environmental treaty secretariats: Stage-hands or actors?', in Fridtjof Nansen Institute *Green Globe Yearbook of International Cooperation on Environment and Development*. Oxford, Oxford University Press, pp17–30

Schermers, H. G. and Blokker, N. M. (1995) *International Institutional Law: Unity with Diversity*. Third revised edition. The Hague, Martinus Nijhoff Publishers

Sebenius, J. (1984) *Negotiating the Law of the Sea*. Cambridge, Harvard University Press

Sjöstedt, G., Spector, B. I. and Zartman, I. W. (1994) 'The dynamics of regime-building negotiations', in Spector, B. I., Sjöstedt, G. and Zartman, I. W. (eds) *Negotiating International Regimes: Lessons Learned from UNCED*. London, Graham and Trotman, pp3–20

Széll, P. (1996) 'Decision making under multilateral environmental agreements'. *Environmental Policy and Law*, vol 26, no 5, pp210–214

Tiempo (1996) *Welcome to the Maldives climate conference*. Issue 22

Underdal, A. (1991) 'International cooperation and political engineering', in Nagel, S. S. (ed) *Global Policy Studies: International Interaction Towards Improving Public Policy*.Basingstoke, Macmillan, pp98–120

Underdal, A. (1994) 'Leadership theory: Rediscovering the arts of management', in Zartman, I. W. *International Multilateral Negotiation: Approaches to the Management of*

Complexity. San Francisco, Jossey-Bass, pp178–200.

Underdal, A. (2002) 'One question, two answers', in Miles, E. L., Underdal, A., Andresen, S. Wettestad, J., Skjaerseth, J. B. and Carlin, E. M. *Environmental Regime Effectiveness: Confronting Theory with Evidence*. Cambridge, MIT Press

UNEP (2003) *Handbook for the International Treaties for the Protection of the Ozone Layer*. Sixth edition, available at www.unep.org/ozone

Wallihan, J. (1998) 'Negotiating to avoid agreement'. *Negotiation Journal*, vol 14, no 3, pp257–268

Weiss, E. B. (1989) *In Fairness of Future Generations: International Law, Common Patrimony and Intergenerational Equity*. Tokyo, United Nations University

Werksman, J. (1996) 'The Conferences of the Parties to Environmental Treaties', in Werksman, J. (ed) *Greening International Institutions*. London, Earthscan

Werksman, J. (1999) *Procedural and institutional aspects of the emerging climate change regime: Do improvised procedures lead to impoverished rules?*. Paper presented at the concluding workshop for the project to enhance policy-making capacity under the Framework Convention on Climate Change and the Kyoto Protocol, London, 17–18 March 1999

Wettestad, J. (1999) *Designing Effective Environmental Regimes: The Key Conditions*. Cheltenham, Edward Elgar Publishing

Winham, G. R. (1977) 'Complexity in international negotiations', in Druckman, D. (ed) *Negotiations: Social-psychological Perspectives*. Beverly Hills, Sage, pp347–366.

Yamin, F. (1998) 'The Kyoto Protocol: Origins, assessment and future challenges'. *Review of European Community and International Environmental Law*, vol 7, no 2, pp113–127

Yamin, F. (2001) 'NGOs and international environmental law: A critical evaluation of their roles and responsibilities'. *Review of European Community and International Environmental Law*, vol 10, no 2, pp149–162

Yamin, F. and Depledge, J. (2004) *The International Climate Change Regime: A Guide to Rules, Institutions and Procedures*. Cambridge, Cambridge University Press

Yefimov, G. K. (1989) 'Developing a global negotiating machinery', in Mautner-Markhof, F. (ed) *Processes of International Negotiations*. Boulder, Westview Press, pp55–64.

Young, O. R. (1991) 'Political leadership and regime formation: On the development of institutions in international society'. *International Organization*, vol 45, no 3, pp281–308

Young, O. R. (1994) *International Governance: Protecting the Environment in a Stateless Society*. Ithaca, Cornell University Press

Young, O. R. (ed) (1997) *Drawing Insights from the Environmental Experience*. Cambridge, MIT Press

Zammit Cutajar, M. (1995) *Submissions from Parties related to AGBM, third session*. Memorandum to Permanent Missions, dated November 1995. On file with secretariat

Zammit Cutajar, M. (1996) *Letter to His Excellency Mr. Chen Chimutengwende, President of COP 2*, 4 March 1996. On file with secretariat

Zammit Cutajar, M. (1999) *Statement at the opening of the high-level segment of COP 5 Bonn, 2 November 1999 by Michael Zammit Cutajar, Executive Secretary, UNFCCC*, available at http://unfccc.int/cop5/media/cop5hls.html

Zartman, I. W. and Berman, M. R. (1982) *The Practical Negotiator*. New Haven/London, Yale University Press

Zartman, I. W. (1994) 'Two's company and more's a crowd: The complexities of multilateral negotiations', in Zartman, I. W. (ed) *International Multilateral Negotiation: Approaches to the Management of Complexity*. San Francisco, Jossey-Bass, pp1–12

Zartman, I. W. (1999) 'Introduction: Negotiating cultures', in Berton, P., Kimura, H. and Zartman, I. W. (eds) *International Negotiation: Actors, Structure/Process, Values*, London, Macmillan, pp1–10

Zartman, I. W. and Rubin, J. Z. (ed) (2000) *Power and Negotiation*. Ann Arbor, University of Michigan Press

Index